*Centre for the History of Music in Britain,
the Empire and the Commonwealth*

CHOMBEC WORKING PAPERS No. 1

The Sounds of Stonehenge

Edited by

Stephen Banfield

BAR British Series 504
2009

Published in 2016 by
BAR Publishing, Oxford

BAR British Series 504
Centre for the History of Music in Britain, the Empire and the Commonwealth
CHOMBEC Working Papers No. 1

The Sounds of Stonehenge

ISBN 978 1 4073 0630 8

© The editors and contributors severally and the Publisher 2009

COVER IMAGE *from Charles Knight:* Old England: a pictorial museum *(London, 1845)*

The authors' moral rights under the 1988 UK Copyright,
Designs and Patents Act are hereby expressly asserted.

All rights reserved. No part of this work may be copied, reproduced, stored,
sold, distributed, scanned, saved in any form of digital format or transmitted
in any form digitally, without the written permission of the Publisher.

BAR Publishing is the trading name of British Archaeological Reports (Oxford) Ltd.
British Archaeological Reports was first incorporated in 1974 to publish the BAR
Series, International and British. In 1992 Hadrian Books Ltd became part of the BAR
group. This volume was originally published by Archaeopress in conjunction with
British Archaeological Reports (Oxford) Ltd / Hadrian Books Ltd, the Series principal
publisher, in 2009. This present volume is published by BAR Publishing, 2016.

Printed in England

BAR titles are available from:

	BAR Publishing
	122 Banbury Rd, Oxford, OX2 7BP, UK
EMAIL	info@barpublishing.com
PHONE	+44 (0)1865 310431
FAX	+44 (0)1865 316916
	www.barpublishing.com

CONTENTS

	page
FOREWORD	ii
LIST OF ABBREVIATIONS	iii
USING THE LIST OF SOURCES	iii
CREDITS	iii
LIST OF CONTRIBUTORS	iv
LIST OF FIGURES	v

PART 1 ARCHAEOACOUSTICS AND BEYOND

1. The sounds of Stonehenge: some notes on an acoustic archaeology
   Joshua Pollard — 1

2. New art—ancient craft: making music for the monuments
   John Crewdson and Aaron Watson — 4

3. Soul music: instruments in an animistic age
   Simon Wyatt — 11

4. Songs of the stones: the acoustics of Stonehenge
   Rupert Till — 17

PART 2 CULTURAL HISTORY

5. The cultural history of Stonehenge
   Ronald Hutton — 43

6. Megaliths in English art music
   Stephen Banfield — 46

7. Stonehenge and its film music
   Guido Heldt — 56

8. Stonehenge in rock
   Timothy Darvill — 66

LIST OF SOURCES — 75

FOREWORD

The Sounds of Stonehenge originated as a workshop of the Centre for the History of Music in Britain, the Empire and the Commonwealth (CHOMBEC), held at the Victoria Rooms, University of Bristol on 28 November 2008. I had ventured on extending CHOMBEC's brief backward in time to prehistory for two reasons. Some months earlier, Thomas Hardy's description of the sounds of Stonehenge, referred to by several of our authors below, had been raised by the literature scholar Mark Asquith at another CHOMBEC study day, this time on Vaughan Williams's Ninth Symphony. At about the same time, our contributor Simon Wyatt had contacted me with a view to teaching the archaeology of music; we met in a cafe where he produced not only a c.v. but some attractive prehistoric drums which he proceeded to demonstrate. Without at the time even having heard of archaeoacoustics, a field of study which came of age with the publication of Chris Scarre and Graeme Lawson's book of that name in 2006, I began to open my eyes towards a rich interdisciplinary line of investigation.

Interdisciplinary it certainly is. In the pages below will be found material pertaining to acoustic physics, anthropology, archaeology, architecture, cognitive psychology, English literature, film studies, history, history of art, media and popular studies, musicology, sociology, and creative composition (which was also present at the workshop in the form of Aaron Watson and John Crewdson's multimedia presentation of music and visuals, beginning the day most impressively but impossible to reproduce here). Streams of thought bleed productively across different chapters and academic viewpoints, as when the idea of music articulating historical time is mentioned by Joshua Pollard, by Rupert Till in his introduction, and by myself in connection with John Ireland's *Mai-Dun*. Indeed, Till's recreation of a possible prehistoric soundscape at Stonehenge almost becomes imaginative literature in its own right. Ronald Hutton's incisive summary of the 'hard' and 'soft' interpretations of Stonehenge is amply demonstrated in the types of film and film music Guido Heldt uncovers. Pollard, Till and Hutton all mention the traffic problem, and whereas Timothy Darvill's chronicle of the Stonehenge pop festivals climaxes with government intervention of a negative sort, it may be hoped that in a small way this book could help persuade the authorities that the extraneous sounds of Stonehenge—the cars and lorries on the nearby A303 and even nearer A344—are currently its greatest threat, and that they must do something radical to remove them so that its intrinsic sound capabilities, so fully explored by Till and others below, can once again reach their potential.

Stephen Banfield
October 2009

LIST OF ABBREVIATIONS

AHRC	Arts and Humanities Research Council
cal	calibrated radiocarbon dating
BP	before the present
bpm	beats per minute
dB	decibel(s)
EDT	early decay time
EPSRC	Engineering and Physical Sciences Research Council
Hz	hertz
kHz	kilohertz
pers comm	personal communication
STi	speech transmission index
TRB	Trichterbecherkultur (Funnel beaker culture)

USING THE LIST OF SOURCES

Author–date citations in the main text give the minimum information necessary for location in the alphabetical sequence: surname, page number where applicable, initial(s) where there is more than one author with the same surname, date where there is more than one entry per author. Any author citation not traceable in the section for books and articles will be in the internet sequence. The abbreviation *et al* is used in the list of sources when there are more than three authors, additionally in the text when there are more than two.

CREDITS

Sir Harrison Birtwistle: *Yan Tan Tethera: a mechanical pastoral.* © 1984 by Universal Edition (London) Ltd, London/UE 17686. Music reproduced by permission.
Roger Hutchinson copyright reserved.
John Ireland: 'Le Catioric', from *Sarnia*. © 1941 by Hawkes & Son (London) Ltd. Reproduced by permission of Boosey & Hawkes Music Publishers Ltd.
John Ireland: *Legend*. © 1938 by Schott & Co Ltd, London. Reproduced by permission of Schott Music Ltd. All rights reserved. International copyright secured.
John Ireland: *Mai-Dun* and *The Forgotten Rite*. Reproduced by permission of Stainer & Bell Ltd, London.
Photographs by John Ireland reproduced by permission of the John Ireland Trust.
Plan of Maryhill/Stonehenge reproduced by permission of Ernie Piini.
John Pickard: *Men of Stone*. Reproduced by permission of Kirklees Music.

Colour illustrations appear on pp. 29–30 and 69–70 and are identified as such in the text.

LIST OF CONTRIBUTORS

Stephen Banfield is professor of music at the University of Bristol and was founding director of CHOMBEC, its Centre for the History of Music in Britain, the Empire and the Commonwealth, from 2006 to 2009. He has published five books on English and American music and his next two will be histories of music and its soundscapes away from the metropolis—around the British Empire and in the west of England.

John Crewdson is a composer and academic currently completing a PhD in behavioural musicology at Royal Holloway University of London. His doctoral research focuses on the drives and values that underpin our musicality. Other interests include investigating the nature of experience at Neolithic and Bronze Age monumental sites and creating audio compositions informed by such investigation.

Timothy Darvill is professor of archaeology in the School of Conservation Science, Bournemouth University. He specialises in the Neolithic of northwest Europe, prehistoric ceramics, and archaeological resource management. In 2008 he directed with Geoffrey Wainwright an excavation inside Stonehenge in order to better understand the date and purpose of the double bluestone circle. He also has fieldwork projects in Pembrokeshire, looking at the source of the Stonehenge bluestones, in the Isle of Man, and in the Cotswolds.

Guido Heldt, lecturer in music at the University of Bristol since 2004, was previously lecturer in music at the Free University, Berlin and visiting lecturer at Wilfrid Laurier University, Waterloo, Ontario. His PhD from the University of Münster on early 20th-century English tone poems was published in 2007. He works currently on film music and narrative theory, on composer biopics and on musical films in Nazi cinema.

Ronald Hutton is professor of history at the University of Bristol and the author of 14 books and scores of essays, many concerned with images of ancient religion in British culture.

Joshua Pollard is reader in archaeology at the University of Bristol. His interest in prehistory and monumentality has led to involvement, as co-director, in major fieldwork projects at Avebury and Stonehenge.

Rupert Till is senior lecturer in music technology at the University of Huddersfield, and is a composer, musicologist and producer. His publications have included writings on popular music, trance and religion, as well as musical recordings and productions. He is primary investigator of the AHRC/EPSRC Acoustics and Music of British Prehistory research cluster.

Aaron Watson is an artist and archaeologist investigating the multisensory qualities of Neolithic and Bronze Age monuments and landscapes, including ongoing acoustic investigations at locations such as Avebury, Stonehenge and Maeshowe. His research organisation and consultancy (Monumental) collaborates with archaeologists, artists, museums and the heritage industry to explore new ways of recording, interpreting and communicating the past through fieldwork, publication, artworks, multimedia, landscape installation and exhibitions.

Simon Wyatt studied archaeology at the University of Edinburgh at undergraduate and postgraduate levels. His interests are prehistory, music, cognition and craft techniques, including the making of model instruments and different styles of basketry. He is currently an honorary research fellow at the University of Bristol.

LIST OF FIGURES

2.1: Live performance by the authors using 24-bit sound and slide projection at the Theoretical Archaeology Group conference, Cardiff University, 1999.

2.2: Sound experiments by the authors inside Maeshowe, Orkney, using electronic synthesis to awaken acoustic effects latent within the format of the architecture.

2.3: A view of Avebury from the top of the substantial earthen bank which interrupts the movement of sound. The Kennet avenue enters the monument through an entrance beside the prominent stand of trees.

2.4: Data collected along a transect leading from the centre of Stonehenge and through the sarsen circle, emphasising the increased volume within the stone circle compared to the control. All measurements are relative to the recorded volume of sound at the source. Plan after Cleal *et al*.

2.5: The outer stone circle at Stonehenge, which acts as a filter, not only obscuring the view into the interior of the monument but significantly obstructing the movement of higher frequency sounds.

2.6: Data collected along a transect around the outside of the sarsen circle at Stonehenge, emphasising how the stones differentially filter high and low frequency sound. The notable trough in the high frequency range between stones 4 and 7 corresponds to the additional presence of stones in the trilithon horseshoe that lie between the listener and the sound source. All the measurements are relative to the recorded volume of sound at the gap between stones 1 and 30. Plan and stone elevations are shown schematically (after Cleal *et al*).

2.7: Looking and listening while walking around the outside of the sarsen circle at Stonehenge. The percussive sound of drumsticks being struck at the centre of the monument becomes detached from the visible act, breaking down cause and effect.

3.1: Prehistoric flutes and whistles from Veyreau (a), Avebury (b), Sewell (c), Charavines (d), West Kennet (e), Falköping (f) and Tiszapolgár-Basatanya grave 67 (g). Images drawn from photographs in Bognár-Kutzián, Clodoré 2002a, Lund 1991 and Piggott 1962; after a drawing in Merewether; and after a photograph by N Wyatt.

3.2: Prehistoric drums from Böhlen (a), Zorbau-Gerstewitz (b), Hornsömmern (c) and Langenburg pit 95 (d). Images redrawn after Behrens and Schröter, Mildenberger 1952, D W Müller and Nitzschke.

4.1: Theoretical consideration of sound reflections in Stonehenge

4.2 (colour): The Maryhill Monument, Maryhill Museum of Art, Washington State, USA (author's photo)

4.3: Reverberation time at Maryhill

4.4: Impulse response of Maryhill

4.5: Resonance at Maryhill

4.6 (colour): Real and theoretical modes at Maryhill

4.7: Reverberation time T30 at different frequencies. Purple is at the entrance to the space by the bluestone ring, grey is central in the entrance a few metres in, red is at the Heel Stone.

4.8: Acoustic calculations for the centre of Stonehenge

4.9: Binaural impulse response

4.10 (colour): Speech transmission index above 0.9 as indicated by black lines

4.11 (colour): LG80 envelopment mapped out in the space

4.12: Comparative acoustic values

4.13: Acoustic values in different measurement (listening) positions (central source at 1kHz, values within suggested range R in bold)

4.14: Results positions P1–P7 for 4.14

4.15: Reflections from a centre source to receiver (right)

6.1: Thomas Lloyd Fowle, 'The Stonehenge polka', sheet music cover

6.2: Fowle, 'The Stonehenge polka', section incorporating 'God save the Queen'

6.3: Luke Cavendish Everett, 'The "Prehistoric"', sheet music cover

6.4: Norman Kennedy, 'Prehistoric zig-zags', sheet music cover

6.5: Photographs of Maiden Castle and Arthur G Miller taken by John Ireland, 1923

6.6: George Lloyd, *Iernin*, poster (1935)

6.7: John Pickard, *Men of Stone*, front cover

6.8: Harrison Birtwistle and Tony Harrison, *Yan Tan Tethera*, Opera Factory production. 'Bad'un' and standing stone in the background

6.9: John Ireland, *Sarnia*, first movement ('Le Catioroc'), extract

6.10: Musical motifs in Harrison Birtwistle and John Ireland

7.1: *Tess of the d'Urbervilles*, BBC mini-series (2008)

7.2: Alan Lisk, *Tess of the d'Urbervilles*, pastoral melody (aural transcription)

7.3: *Tess of the d'Urbervilles*, London Weekend Television (1998), close-up

7.4: *Tess of the d'Urbervilles*, London Weekend Television (1998)

7.5: Alan Lisk, *Tess of the d'Urbervilles*, 'memory' theme (aural transcription)

7.6: David Milner ed, *The Highways and Byways of Britain*, cover

7.7: *Shanghai Knights* (2003)

7.8: Randy Edelman, *Shanghai Knights*, pastoral melody (aural transcription)

7.9: *The Tomb of Ligeia* (1964)

7.10: Nic Rowley and Marc Wilkinson, *Quatermass*, chorale (aural transcription)

7.11: *Quatermass* (1979)

7.12: *Night of the Demon* (1957), opening

7.13: Clifton Parker, *Night of the Demon*, chromatic motif (aural transcription)

7.14: *Muppets from Space* (1999)

7.15: *Evil Aliens* (2005)

7.16: *Help!* (1965), Stonehenge in the distance

7.17: *Help!* (1965), aerial shot

8.1: Stonehenge, Wiltshire. View of the central stone settings looking southwest (author's photo)

8.2 (colour): Stonehenge Rocks! Sew-on clothing patch from the range of English Heritage merchandise introduced in 2008 (author's photo)

8.3 (colour): Ten Years After, *Stonedhenge* (1968)

8.4 (colour): Yes, *Tales from Topographic Oceans* (1973). Inner cover artwork by Roger Dean.

8.5 (colour): Magazine advertisement for the Sonic Rock 2005 festival featuring Stonehenge and a flying saucer. Reproduced from promotional artwork

8.6: Poster designed by Roger Hutchinson advertising the 1975 Free Festival at Stonehenge.

8.7: Aerial view of Stonehenge with the 1984 festival in full swing

PART I

ARCHAEOACOUSTICS AND BEYOND

CHAPTER 1

The sounds of Stonehenge: some notes on an acoustic archaeology

Joshua Pollard

What are the possibilities for an archaeology of sound at Stonehenge? While the final form of the stone monument invested it with certain acoustic properties, whether these were intended or not is open to debate. In fact, can we even be sure that the monument was an arena in which orchestrated sounds (music, chant, speech) were emitted? And how do we explore these issues? Such points acknowledged, in this brief contribution it is suggested that we should not constrain or deny interpretive possibilities. Sound was and is a part of Stonehenge. We can be sure of the importance of sound in the sensory geographies of the many generations who constructed and encountered the monument, and so must think of how an acoustic archaeology of Stonehenge can be taken forward.

While few would doubt that Stonehenge is an exceptional prehistoric monument, the question must still be asked, 'Why is its fame so great?' There are other, broadly contemporary constructions of comparable magnitude, such as the great passage graves of Knowth, Dowth and Newgrange in the Boyne valley of Ireland, or the megalithic alignments at Carnac in Brittany, which rank equal if not greater in scale but do not command the worldwide currency of Stonehenge. When it comes to the fame of Stonehenge, matters of form, history and imagination can all be seen to contribute. Its complex construction made it evident as an 'artificial' work during times when other megalithic monuments and barrow mounds were commonly relegated to the domains of nature and the supernatural. Since the 17th century and the work of early scholars such as Inigo Jones, John Aubrey and William Stukeley, the monument has been under a continuous academic gaze, generating a wealth of interpretations on its age, purpose and the identity of those responsible for its creation (see Chippindale 3/2004 for full details). Reflecting an interpretive inclusivity that has often attached itself to megalithic monuments, lay and 'alternative' theories have flourished in tandem, a percentage being sensible, others reflecting contemporary obsessions with themes such as ancient astronomy, lost religious traditions, alien visitations and earth energies. However, it is not just the appeal of megalithic mystery that has made the site the cultural phenomenon it is today. Since the Enlightenment, Stonehenge has been emblematic of the qualities of antiquity, ruin and the sublime—a place to meditate on the passing of time, of memory and personal and cultural atrophy. Arguably, it is the ruinous state, though one that is not so great as to prevent an appreciation of original form, that gives the monument much of its popular and academic appeal.

Above all, Stonehenge is a monument of interpretive fecundity, and even in its ruinous state it seems to offer the potential to understand the experiences and performances that characterised its original use. Through an engagement with its physical form and that of the surrounding landscape, various phenomenological studies have attempted to replicate the experience of the monument as it is approached along the earthwork avenue, and the fields of view into and out of the stone settings (Cleal *et al*; Tilley *et al*). Beyond ocular experience, Aaron Watson (2006; Watson and Keating 1999; this volume) and Rupert Till (this volume) have explored the complex soundscapes produced by the monument's architecture. These studies have revealed how noise is variously contained, displaced and resonated by the megalithic settings, generating some curious effects in the process. However, as Scarre (2006) and Watson (2006) have highlighted, simply because an architectural form possesses dramatic and unusual acoustical properties need not automatically imply that it was created with such in mind: much could be incidental and unintended.

Demonstrating acoustic intentionality will always be problematic, especially in a prehistoric context, and we are left wondering how important sound and music might have been in the various ceremonies conducted at the monument. This should not, however, generate too much pessimism over inferential limits. Even though it is unlikely that the megalithic architecture of Stonehenge was designed with sound in mind, the fact that it could produce some remarkable acoustic effects is itself of considerable interest. First, we have to think of how such effects would have been comprehended by the Neolithic and Bronze Age communities who built and engaged with the monument, and this must involve consideration of emic understandings of causation and distributed agency of both human and non-human form. In animistic mode, acoustic effects might therefore be seen as evidence of miraculous or supernatural qualities inherent within the place itself or in the stones—the structure 'acting back' (Tuzin)—or deriving from supernatural agencies that inhabited the monument: 'an echo might be the voice of an ancestor rather than a sound wave' (Watson 2006: 20). Sound could have added considerably to the potent qualities and efficacy of the monument and sense of the numinous.

Second, if we acknowledge that music (be it simple repetitive drumming, chanting, singing, or more sophisticated performances) is a recurrent if not universal feature of ritual performance, then we might reliably infer that orchestrated, or at least intentional, sound played a part in the life of a ceremonial monument such as Stonehenge. (From the ceremonies of new order Druids, to war memorial services and the now defunct Stonehenge Festival, music has certainly shaped more recent performances.) Admittedly, direct archaeological evidence for musical practice during the British late

Neolithic—the time of Stonehenge—is remarkably scarce; but to think of a Neolithic world without music seems inconceivable. Convincing instruments are absent from the record, though large Grooved Ware vessels, including those from the contemporary site at Durrington Walls to the east of Stonehenge, could, if tightened skins were fixed across their mouths, have acted as very effective drums (see chapter 3 below). In fact, with applied cordons and 'stitch-like' incised decoration, the form of some of the Grooved Ware vessels from Durrington might easily be interpreted as skeuomorphic or organic drums (see for example Wainwright, fig. 33). Other objects of antler, wood, and stone are possible contenders for percussive instruments; and then there is the human body itself through which singing, nasal noise, clapping, stamping and other sounds could be performed. Just because the sounds of Neolithic music are long gone, not directly recoverable, this should not be taken as indicative of their unimportance. Music was surely an essential and deeply embedded component of cultural life during the period, and one through which a sense of identity and connection with past generations and other worlds was performed.

Those studies of the acoustic properties of Stonehenge undertaken by Till and Watson have focused on the monument in its most developed and familiar form, as it existed during the second half of the 3rd millennium BC (the latest Neolithic and earliest Bronze Age). They have also approached the site as a completed monument, a state which it may never have achieved. Through much of its active life Stonehenge was a monument in motion, undergoing various episodes of construction and modification that spanned the earliest 3rd to the mid-2nd millennium BC (see Cleal *et al* for the detailed sequence). Even after the last constructional phase it did not remain static, undergoing a process of decay and depredation in an ever changing landscape (Darvill 2006; A J Lawson), resulting in the Stonehenge we see today. Transformations to the monument and the continual inhabitation of the surrounding landscape since prehistory have produced a changing soundscape, one that is perhaps more varied than those of many other locations in the British Isles. The possibility therefore exists to undertake a diachronic archaeology of sound, a kind of cultural history of noise, for Stonehenge. Whether this could be put to musical score remains to be seen (though see Banfield, chapter 6 below, on another prehistoric site, Maiden Castle, as a musical work). Some brief thoughts are here offered.

The acoustic history of Stonehenge would, with some variation, be one of progressive increase in the variety and level of sound. There are notable points of contrast. For much of its 'middle age', during the later prehistoric and early historic periods, when little visited or even avoided, the ambient sound of the monument was of animal and bird life and the rhythms of agrarian activity emanating from the surrounding landscape. The romantically-inspired depictions of the monument by Turner, Girtin and Constable set to perpetuate this image of the remote, ruined monument, visited only by sheep and shepherd (and the occasional lightning strike), yet even by the early 19th century Stonehenge had become a busy tourist attraction. There was the clatter of carriages, later bicycles, surreptitious hammering as visitors attempted to remove pieces of 'souvenir' stone, and the exuberant sounds of periodic fairs, hare coursing and cricket matches (see Chippindale 3/2004 for details). Even points of relative silence were broken, as in 1797 and 1900, by the dull resonating thud of a falling stone.

Today, the soundscape is predominantly of visitors and vehicles. Within the monument it has its ebbs and flows, but from the middle distance comes the omnipresent rumble of traffic along the A303—an acoustic as well as visual pollution to counter which there have been several unsuccessful proposals for the removal of the road or its placement within a bored tunnel. In the 21st-century world of Stonehenge, sound has become politics. Curiously, given Stonehenge's emblematic status as a symbol of antiquity and primitive technology, the recent history of the monument is closely embroiled with many other forms of modern motive technology. The progressive militarisation of Salisbury Plain since the late 19th century has brought to the site the sound of tanks, planes and helicopters, as well as distant heavy gunfire. Between 1916 and 1920 a military airfield was even in operation on Stonehenge Down, immediately to the west of the monument.

It goes without saying that very different soundscapes accompanied the early life of Stonehenge. In its first phase, constructed *ca*2950 BC, it was an open monument, defined by a circular ditched enclosure enclosing a ring of 56 pits or stoneholes, possibly for the Welsh bluestones (Cleal *et al*; Parker Pearson 2007, 2009). Sound would have travelled freely within and without the monument. During the middle of the 3rd millennium BC a radical remodelling took place, involving the creation of the sarsen outer circle and central horseshoe of trilithons, replacing a slightly earlier ring of bluestones. This configuration, which may have drawn upon the symbolism of the house (Pollard 2009), produced a closed format with very different acoustic qualities (as described by Till and Watson below). But what kinds of activities took place within the monument? From the beginning Stonehenge was associated with the burial and veneration of the dead, indicated by the large number of cremation burials and disarticulated human bones placed within the ditch, the tops of the Aubrey Holes and the periphery of the interior. Modern sensibilities would suggest that this, one of the largest cemeteries, if not the largest, in late Neolithic Britain, may have attracted quiet veneration. However, ethnographic instance tells of how in many traditional societies, noisy ceremonies accompany the veneration of the dead and ancestors.

To focus upon the acoustic qualities of Stonehenge as a completed monument is to ignore important aspects of its experiential qualities that might be derived instead from a 'construction perspective'. The ancestral spirits residing at Stonehenge were also disturbed, or quietly invigorated, by the sounds of construction. Creating the megalithic settings generated not only the social noise that accompanies large gatherings and the repetitive and penetrating sound of stone on stone as the bluestone and sarsen megaliths were shaped by flaking and pounding, but also the shouting and exertion released during the moving of earth, timber and stone, which could have been co-ordinated into call-and-response vocalisation of the sort that in some societies produces work songs. Such working was neither an easy nor a rapid task, and it is likely that during the mid- to late 3rd millennium BC the acoustic landscape of construction more frequently characterised Stonehenge than that of formal ceremony. How such sounds were understood—whether auspicious or simply neutral—is difficult to know, but they provided a distinctive acoustic texture that contrasted with the more familiar sounds of sociality (gathering, feasting, the rhythms of daily life) that characterised contemporary activity on the Avon riverside at Durrington Walls.

Past sound is not directly recoverable, but working with the physical fabric of past activity always offers potential for its reconstruction and interpretation. The brief thoughts offered here have hopefully illustrated the possibility of creating a more detailed acoustic history of Stonehenge. Context, of course, remains everything and in interpretation such studies should always seek to understand the particular, localised and emic aspects of causation and acoustic associations. Sound, however understood, occasionally in the form we might describe as music, was always a part of Stonehenge, and it merits study that is equal to the ocular and haptic experience that has to date characterised archaeological phenomenology.

NOTE
Stonehenge 1 was not the first monumental construction in this region. However, its form was novel by comparison with the earthen long barrows, two cursus monuments (massive linear enclosures) and causewayed enclosure at Robin Hood's Ball created during the earlier Neolithic (the 4th millennium BC).

CHAPTER 2

New art—ancient craft: making music for the monuments

John Crewdson and Aaron Watson

FIGURE 2.1: LIVE PERFORMANCE BY THE AUTHORS USING 24-BIT SOUND AND SLIDE PROJECTION AT THE THEORETICAL ARCHAEOLOGY GROUP CONFERENCE, CARDIFF UNIVERSITY, 1999.

Introduction

Neolithic and Bronze Age monuments stand silent in the modern world, yet there is a possibility that structures such as Stonehenge and Avebury were once filled with a rich variety of expressive sound. Archaeology has long interpreted monuments as venues for social gatherings, and anthropological studies suggest that such events are often accompanied by diverse sound and music-making (Merriam; Blacking 1973). Archaeoacoustic research has demonstrated that ancient architecture can influence the behaviour of sound in distinctive ways, irrespective of whether such effects were intentional (Crewdson; Devereux and Jahn; Watson 2006; Watson and Keating 1999, 2000). Monuments were capable of hosting complex ritual activities, and music-making is a common component of ritual practice (Blacking 1973; Blumenfeld; Hargreaves and North; Nketia). Sometimes the sounds produced are believed to be the voices of gods or ancestors or believed to summon, control or placate supernatural forces (Tuzin). The rhythmic and emotive properties of music also allow large group-actions to be easily controlled (Bloch; Jackson; Schafer 1993: 31), with the mood of the music communicating the meaning of the ritual and indicating the correct way, and when, to respond.

The enormous potential for these buildings to stage theatrical multisensory events has compelled the authors to explore their auditory possibilities through the creation of artistic audio-visual works (e.g. Watson and Was). The diverse possibilities for visually representing the experience of monuments have been considered elsewhere (Watson 2004), and in this chapter we focus upon sound and music. The creation of these present-day experiences supersedes the sensory limitations of traditional printed media, and allows an audience to be immersed within soundscapes which are both heard *and* felt (Fig. 2.1). It is not our aim to create 'reconstructions' of ancient practice and experience. Rather, this is an experimental and contemporary approach to interpreting ephemeral qualities of the archaeological record; an attempt to create new art from the residue of ancient craft.

Ancient sound and music
Sound was fundamental to the lives of people in the Neolithic and Bronze Age. The natural soundscape was a *hi-fidelity* audio environment (Schafer 1994), rich in texture with complex dynamics provided by the natural world and its inhabitants. 'Natural' sounds would have been of great significance, indicating location, the presence of animals, the time of day, the seasons, warnings of danger, and even a means for climatic and environmental prediction, just as in non-western societies today they can be central to navigation, language and music (e.g. Feld 1982, 1996; Gell 1995; Kawada; Pocock 1989). In addition to natural sound, an important aspect of audition would have arisen in the absence of the written word. Aural communication would have had a heightened significance, encouraging the development of complex symbolic and abstract auditory associations and a refined auditory memory (Botha and Knight). Complex spoken 'compositions' such as story-telling would have been essential for memorising cultural histories and would have also played a central role in defining individual identity (Stokes). Sound was, and remains, integral to the identity of places and landscapes (Pocock 1988; Seeger 1994; Tuan 1974).

There must also be a consideration of the deliberate artistic organisation of sound into music. While debate about the earliest origins of music and instrumentation continues (see Zubrow and Blake), it has been suggested that there is 'unambiguous evidence of musical behaviours' dating back to *ca*150,000 BP (Cross and Morley: 77). Upper Palaeolithic cave art dating between 10,000 and 35,000 BP might have had an acoustic dimension, with images preferentially located in resonant caverns (Davois; Reznikoff; Reznikoff and Dauvois). Impact marks upon stalactites suggest a knowledge and use of pitched sound, as these features resonate when struck (Dams). Numerous bird bone artefacts with 'finger holes' are believed to be musical pipes capable of pitched melodies. Two such 'pipes' found at Geissenklösterle in Germany show that these artefacts were already well established around 36,000 BP (Hahn). Additional fragmentary remains of twenty bird bone pipes with finger holes were found in the Grotte d'Isturitz in southern France. G Lawson *et al* suggest that these pipes are 'unequivocally musical and have a sophisticated design' (113). The technology to produce pitched sounds in the Palaeolithic suggests a culture engaged in music-making. There is little doubt that 'music' existed in the Neolithic, although maybe not in a form we would readily recognise today. Although there are few examples of unambiguous sound-producing devices from the British Neolithic and Bronze Age (see Lund 1981; Megaw 1960, 1968, 1984), we infer that people were likely to have been competent, if not sophisticated, musical practitioners.

If music-making had a role in Neolithic society, how might this music have sounded? Some insight can be gained by examining the surviving instrumentation elsewhere in Europe. A study of Scandinavian Neolithic organology has revealed diverse artefacts which have potential as 'sound-producing devices' (Lund 1981: 246). Rattles can potentially be made out of animal teeth, shells, and other bone objects, while any serrated surface could have acted as a scraper (Lund 1991: 38). Percussive instruments include wooden logs, sticks, nutshells, vessels of clay and other bone objects. Artefacts containing chambers or cavities can be used as resonators to amplify various qualities of the voice or other instruments. Bows could have been used as musical instruments. An especially striking instrument was the bull-roarer, an example of which has been excavated from a Mesolithic context in Denmark (Lund 1984: 256–7). These instruments typically consist of a stone or heavy object tied to a cord which is then swung to create a striking whirring sound. Pitched instruments available to Neolithic musicians would have included pipes, flutes and cow horns (Lund 1991: 42). Pipes were constructed from bone or reed and were probably end-blown like a bottle. Even struck flint flakes can have distinctive sound properties (Cross and Watson: 113–14), and a modern flint-knapper can judge raw material quality by listening to its sound (Lund 1991: 38).

Monumental evidence
The soundworld would have changed significantly in the Neolithic with the construction of artificial earth and stone monuments on a large scale. The internal environments of these places possessed unique acoustic qualities which could not have been reproduced elsewhere in the landscape, and might even be considered as instruments in their own right.

Enclosed stone chambers such as the West Kennet long barrow in Wiltshire, or Maeshowe in Orkney, offered near silent environments within which entirely new worlds of sound could be created (Watson and Keating 1999; Watson 2001a), while the rather more open spaces of stone circles also have effects (Crewdson; Watson 2001b; Watson and Keating 2000). In these conditions, instruments would have been heard with a new clarity, possibly leading to their refinement and to an improved knowledge of acoustic phenomena such as the overtone series. Interiors also produce a large dynamic range allowing very quiet sounds to be heard as well as the production of very loud sound due to the highly reflective surfaces and low absorption of stone walling (Fig. 2.2). In the same way that a modern musician is inspired in composition and performance by the use of a new instrument, it is possible that monumental acoustics kindled a surge in creative exploration by the music and instrument makers of the time.

The Sounds of Stonehenge

FIGURE 2.2: SOUND EXPERIMENTS BY THE AUTHORS INSIDE MAESHOWE, ORKNEY, USING ELECTRONIC SYNTHESIS TO AWAKEN ACOUSTIC EFFECTS LATENT WITHIN THE FORMAT OF THE ARCHITECTURE.

FIGURE 2.3: A VIEW OF AVEBURY FROM THE TOP OF THE SUBSTANTIAL EARTHEN BANK WHICH INTERRUPTS THE MOVEMENT OF SOUND. THE KENNET AVENUE ENTERS THE MONUMENT THROUGH AN ENTRANCE BESIDE THE PROMINENT STAND OF TREES.

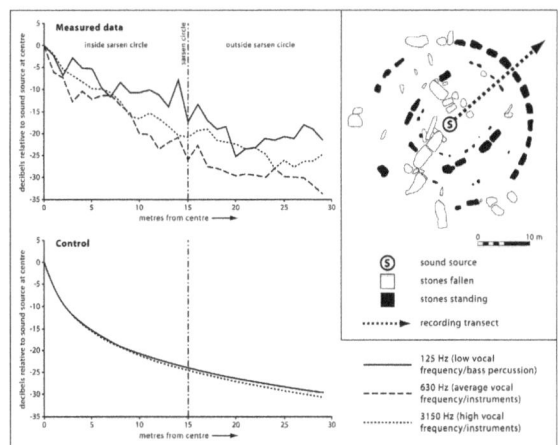

FIGURE 2.4: DATA COLLECTED ALONG A TRANSECT LEADING FROM THE CENTRE OF STONEHENGE AND THROUGH THE SARSEN CIRCLE, EMPHASISING THE INCREASED VOLUME WITHIN THE STONE CIRCLE COMPARED TO THE CONTROL. ALL MEASUREMENTS ARE RELATIVE TO THE RECORDED VOLUME OF SOUND AT THE SOURCE. PLAN AFTER CLEAL *ET AL*.

FIGURE 2.5: THE OUTER STONE CIRCLE AT STONEHENGE, WHICH ACTS AS A FILTER, NOT ONLY OBSCURING THE VIEW INTO THE INTERIOR OF THE MONUMENT BUT SIGNIFICANTLY OBSTRUCTING THE MOVEMENT OF HIGHER FREQUENCY SOUNDS.

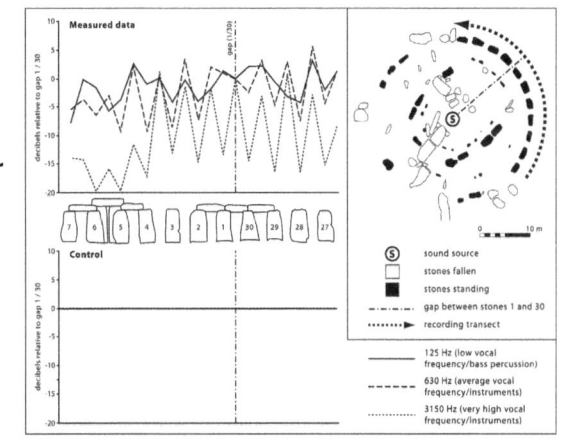

FIGURE 2.6: DATA COLLECTED ALONG A TRANSECT AROUND THE OUTSIDE OF THE SARSEN CIRCLE AT STONEHENGE, EMPHASISING HOW THE STONES DIFFERENTIALLY FILTER HIGH AND LOW FREQUENCY SOUND. THE NOTABLE TROUGH IN THE HIGH FREQUENCY RANGE BETWEEN STONES 4 AND 7 CORRESPONDS TO THE ADDITIONAL PRESENCE OF STONES IN THE TRILITHON HORSESHOE THAT LIE BETWEEN THE LISTENER AND THE SOUND SOURCE. ALL THE MEASUREMENTS ARE RELATIVE TO THE RECORDED VOLUME OF SOUND AT THE GAP BETWEEN STONES 1 AND 30. PLAN AND STONE ELEVATIONS ARE SHOWN SCHEMATICALLY (AFTER CLEAL *ET AL*).

Fieldwork undertaken by the authors at Stonehenge and Avebury supports the idea that monuments were able to provide extraordinary sensory experiences.

Avebury

A striking 'theatrical' effect could have been encountered by people entering or leaving the Avebury henge monument. The large ditch and bank were acoustically profiled as part of a larger study into the acoustic properties of the monument (Crewdson). The most obvious role of the earthwork and ditch, acoustically speaking, is to contain sound energy created within the monument, effectively isolating the interior (Fig. 2.3). As would be expected, this substantial earthwork acts as an efficient sound barrier that blocks out all sound above 10kHz. Results showed that sound with frequencies ranging between 450Hz to 10kHz were attenuated by an average of 40dB, and for frequencies under 450Hz there was an average 20dB drop. This means that most sounds produced within the monument are shielded by the bank and therefore largely inaudible to listeners outside. This soundproofing capability could have been used to dramatic effect for those approaching and entering the monument along the West Kennet avenue. The stone-lined avenue does not lead directly into one of the entrances; instead, as it approaches the ditch, it starts to veer away from the monument as if to pass it by altogether on its western flank. This takes it away from the line of sight of the south entrance. The avenue then suddenly changes direction again, veering to afford the first clear view of the interior of the monument framed through the southern entrance and between two of the largest standing stones at the site. The effect of suddenly perceiving the interior of Avebury is a striking visual experience (Barrett; Thomas), but would also have had a significant auditory impact. In the short final approach, sounds emerging from the monument would rise in intensity by about 40dB: a large crescendo peaking at the moment when the interior of the monument was visually revealed for the first time. It is compelling to interpret these features as a theatrical audio-visual device.

Stonehenge

The complex stone arrangements we see at Stonehenge today represent the third phase of construction at the monument, *ca*2500–1600 BC (Cleal *et al*). Archaeoacoustic investigations conducted with acoustician David Keating have explored sound pressure and frequency around the best preserved sectors of the monument, with a sound source placed in the centre (Watson 2006).

Fig. 2.4 shows how the relative amplitude of three sound frequencies varies at one-metre intervals along a line leading from the central loudspeaker out through the outer sarsen circle and towards the Heel Stone. The surviving stones of Stonehenge very effectively contain higher frequency sounds within the interior, and amplify them, suggesting the effect would have been further emphasised when the monument was complete. A series of distinctive peaks and troughs on the graph indicates standing waves, caused by constructive and destructive interference as sound waves are reflected between the large stones. This can produce unnerving effects as sounds can become detached from their source, change in amplitude and pitch and behave in counter-intuitive ways such as becoming quieter as a sound source is approached (Watson 2001a: 186; Watson and Keating 1999: 329–30).

Similar to the bank at Avebury, the outer sarsen circle of Stonehenge performs an acoustic role by containing higher frequency sounds (Fig. 2.5). The broad stones of the outer sarsen circle abruptly attenuate higher frequencies, while lower frequencies pass around these sarsens and continue to travel for some distance. The strength of this effect depends upon the location of the listener relative to the gaps between the stones. This is most clearly demonstrated by measurements taken around the perimeter of the northeastern sector of the outer sarsen circle (Fig. 2.6), clearly showing how higher frequencies emerge through gaps between the stones.

Overall, these results emphasise the contrasting acoustic experiences between a listener inside or outside Stonehenge. Within, sound is amplified and enhanced, and there are special effects such as standing waves and echoes. In contrast, sounds emerging into the outside world are both quieter and distorted. Together, these effects would have made the act of moving towards and entering Stonehenge a theatrical transition from ambient sound to sound controlled and contained by the architecture of the monument itself. For a listener approaching Stonehenge when rhythmic drumming and pitched sounds were being produced from within, the experience would not be dissimilar to a modern club DJ mixing dance music; the effect would be like opening up a 'filter' on a dance record, a common mixing technique used to heighten the musical tension before an explosive release for dancers.

Research undertaken at Stonehenge by the authors has also demonstrated the action of early reflections upon speech. Results indicated three strong early reflections with delay times of 0.025sec, 0.033sec and 0.042sec. They each produced a similar spectral response, with the largest peak at 6.1kHz followed by peaks at 4.6kHz, 3.2kHz, 2kHz, 1kHz and a smaller peak at 7.7kHz. The overall effect was an enhanced clarity of voice, the early reflections reinforcing the original voice with delayed versions and adding 'presence' in the form of an upper middle-range frequency boost. In some contrast to the austere, overpowering, even claustrophobic experience of the surrounding stones, voices in the interior are surprisingly strong, bright, and airy.

Another effect may arise from the gaps between the large sarsen stones that make up the outer circle. These could act acoustically as a large 'comb filter'. A comb filter adds a delayed version of the live sound source to itself. Combining the two signals creates constructive and destructive interference patterns which give rise to a range of acoustic effects, including echoes, flanging and standing waves. An interesting effect is produced when white noise is played through a comb filter, where the resulting wave interference pattern can lead to the perception of a residual pitch, non-existent in the original noise signal. This can be quite significant, since pitched sounds are strongly correlated with animate sound sources (Kruth and Stobart). The effect of the comb filter can therefore endow noise-producing, inanimate objects with pitched, animate qualities.

An unexpected experience was observed by a listener standing between stones 1 and 30 (see Fig. 2.6), facing towards the centre of Stonehenge where a second person was striking wooden drum sticks together in a regular rhythm. Intriguingly, the sound appeared to come predominantly from the extreme left and right of the observer rather than the direction of the drum-sticks. In addition, the timing of these disembodied sounds did not coincide with the visual timing of the sticks being struck. This contrived both to create an illusion of disconnection between sound and vision, and also for the production of sound in locations where no sound producing sources were visible. The complex reflections within Stonehenge mean that reflected sound, disconnected in time and space from the event producing it, distorts the listener's expectations of cause and effect (Fig. 2.7). Even to a modern observer, who can rationalise these phenomena through a scientific knowledge of reflections and echoes, the experience is compelling.

Evidence, interpretation and creative composition
Artistic audio-visual works created by the authors are inspired and informed by a combination of empirical evidence, extrapolation, experimentation, and theoretical interpretation. For a musical composition, work begins with the collection of a sound palette of Neolithic instruments via recordings of reconstructed or newly-built instruments, and raw materials collected in the landscape. The next stage involves the reconstruction of the acoustic environments of the monuments. Where possible we incorporate recordings made within the monuments themselves, but this is not always possible due to levels of preservation. Otherwise, acoustic measurements collected from a particular site are combined with calculations derived from survey plans and elevations, sometimes requiring an approximate reconstruction of the original format of the building. In the studio, the data underpinning these essential acoustic properties can be input into digital signal processors and applied to recreate the effects of both stationary and moving sound sources. This combination of a palette of instrumental sound sources and the acoustic environments in which to 'play' them can result in the creation of powerful auditory experiences.

The next stage is to decide how these instrumental sounds can be arranged into audio passages so as to provide an extended, cohesive musical composition. There are certain empirical features that can indicate particular characteristics. One example is that the tempi used in the composition can be predicated on echo-times of a particular site, so that each beat coincides with or is proportional to the time it takes for the sound to be reflected. For example, the outer stone circle at Avebury has a diameter of approximately 330 metres. Were a sound source positioned in the middle of the monument and reflected off the outer ring, the echo would return to the source delayed by 0.97 seconds. To play in time with this delay requires a tempo of around 62 beats per minute, or multiples thereof. Other empirical data can be extrapolated to suggest the dominant pitch in a composition. Natural resonant features, particularly in enclosed spaces like tombs, will reinforce particular frequencies, and since frequency is the perceptual correlate of pitch, the tomb space will in effect be tuned to a certain note or possess a signature timbre. This information, in conjunction with the limited key range of the instrumentation, could be used to indicate a tessitura or even the tonic note of a tonal composition. In addition to tempo and pitch characteristics, certain loudness dynamics can also be predicated on monumental evidence. Earlier in this chapter, the effects of the bank at Avebury were described as effectively creating a natural crescendo in sound for people approaching the monument. By using the acoustic profile data of the ditch and bank (Crewdson) and combining it with a given walking velocity, a crescendo (dB per second) can be calculated and used as a template for crescendi and diminuendi within the composition. While we cannot demonstrate that these pitches, tempi or dynamics would have informed auditory events in Neolithic 'compositions', it is very likely that Neolithic people would have been confronted by these specific frequencies, timings and dynamics.

Sensitivity to the natural soundscape and the heightened significance of aural communication are two factors that might also have influenced acts of composition. Certain features of environmental sounds, like the rhythmic ebb and flow of the wind blowing, rain falling, or the rustle of leaves in trees, could have been used as a guide for the articulation and phrasing of instrumental playing. Furthermore, when moving through the landscape, different natural sound textures intermix, one superseding another. These natural sound transitions could inspire correspondent melodic or thematic transitions in musical composition. Such qualities might even have become enshrined within architecture. For example, the course of the West Kennet avenue at Avebury 'orchestrates' the movements of participants so that they encounter a sequence of contrasting environments along the route, including a river valley. The same is true of the Stonehenge avenue which leads from the banks of the River Avon onto the high ground of Salisbury Plain.

Pitched instrumental sounds could also have been accompanied by human mimicry of sounds of the natural flora and fauna. It is possible that this hybrid orchestration of instrumental music and natural sound would have been accompanied with vocalised themes, organised in an overarching narrative structure. Particular musical features could have overtly represented and evoked specific events and places portrayed in the vocal narrative. If monuments were venues for gatherings of people, then it is likely that they were also locations within which diverse ideas, artefacts, materials, stories, sounds and music were brought together. At large monuments such as Stonehenge or Avebury, or even different chambers within cairns, a variety of activities could have occurred simultaneously, each with its own musical accompaniment. This allows for the possibility of simultaneous thematic counterpoints being experienced and even rudimentary harmony created by chance.

Closing remarks

As a dialogue between archaeological evidence and creative composition, 'artistic' performance offers a new means of engaging with ancient monuments. While we make no claim that these experiences authentically reconstruct Neolithic and Bronze Age music, the silence which presently pervades the past is in its own way no less prescriptive. Fieldwork in support of creative musical composition requires that we pay attention to aspects of sites which would be overlooked by traditional archaeological fieldwork. Multimedia performance is a vehicle for exploring and expressing the sensory and emotive qualities of monuments and landscapes which are difficult to convey in any other way and would otherwise be entirely absent from tellings of the past.

The creation of new art from the residue of ancient craft offers an enriched telling of the past. In contrast to the silent edifices encountered today, megalithic monuments might instead be re-imagined as vibrant auditoria.

ACKNOWLEDGEMENTS

The authors wish to record many thanks to David Keating for continued help and advice with acoustic research and analysis, and to National Trust staff at Avebury for their assistance, and thanks also to Historic Scotland, and in particular the staff at Maeshowe. Research at Stonehenge was made possible in association with English Heritage, Yorkshire Television and BBC Science Radio. We offer many thanks to Nicholas Cook for his patience and guidance, and special thanks to Harry Crewdson and Don Campbell, whose knowledge and support have been invaluable. All images accompanying this chapter were prepared by the authors.

FIGURE 2.7: LOOKING AND LISTENING WHILE WALKING AROUND THE OUTSIDE OF THE SARSEN CIRCLE AT STONEHENGE. THE PERCUSSIVE SOUND OF DRUMSTICKS BEING STRUCK AT THE CENTRE OF THE MONUMENT BECOMES DETACHED FROM THE VISIBLE ACT, BREAKING DOWN CAUSE AND EFFECT.

CHAPTER 3

Soul music: instruments in an animistic age

Simon Wyatt

Preamble

Music is found in every region on earth; it plays a part in shaping our perception of the world. It is constructed and used differently in different cultures, so Handel's use of the key of F signified the pastoral idyll (Blacking 1973), while a Saame *joik*, a traditional song, is specific to an individual, whether human, tree, river or mountain.

We currently know some very early examples of flutes and possibly reeded pipes, which demonstrate the great importance and antiquity of music where our ancestors are concerned. At Geissenklösterle three flutes, one made from hollowed mammoth tusk and two from swan bone, were discovered in recent excavation, dating from *ca*36,000 BC (Conard), while at Isturitz, France, more than twenty largely fragmentary examples were made from vulture bone and date between 20,000 and 35,000 years ago (Buisson; d'Errico *et al*).

We shall take music here to mean 'humanly organised sound', after Blacking (1973). And we shall consider some views on the function and meaning of music, and take note that in prehistory, as we shall see, sound may have had an altogether more metaphorical and even metaphysical significance. Finally we shall acquaint ourselves with the range of instruments that may have been available to the inhabitants at the time of Stonehenge.

The role of music and the self

There is some lateralisation of musical function within the human brain (Falk; Mithen; Morley 2003; Wallin), but it is not as clear-cut as the common view that music resides in the right hemisphere and language in the left. What we may say is that emotional pitch and tone and holistic processing are right hemisphere processes, while semantic, analytical and rhythmic functions are left hemisphere processes.

A parallel division may be seen in the location of the self. So we may state that the self, that is the areas of the brain which correspond to functions related to selfhood or ego, are located in the left hemisphere and are closely related to the language centres (Clark 2006; Jaynes 1990; Ramachandran 1999). When the left hemisphere is unable to maintain the stability of the self, the holistic right hemisphere sustains the equilibrium. Our-'self' and its perception of its place in a group is a necessity of maintaining mental stability. On a larger scale, a society needs to maintain stable relationships with other communities and the cosmos.

Storr, citing D E Berlyne, observes that 'pattern making, gestalt perception, is an integral part of our adaptation. Without it, we should only experience chaos. The creating and perceiving of apprehensible schemes goes on at every conceivable level in our mental hierarchy, from the simplest auditory and visual perceptions to the creation of new models of the universe, philosophies, belief systems, and great works of art including music' (168). Music is a method of pattern making *par excellence*. It is also the result of rhythmic movement, and it may be clearer to think of musical activity as a form of dance. Blacking proposes the use of the term 'bio-social dance' to signify repetitive, patterned, behaviour which distinguishes man from animals. In this repetitive patterned behaviour we may recognise ceremonial activity (Blacking; Donald). The Latin *ritus* means ceremony (Smith: 532), but its Sanskrit cognate *r̥tu*, a rite, implies the perpetuation of *r̥ta,* cosmic order (N Wyatt, pers comm).

Dunbar has proposed that grooming evolved through a musical stage of communication to language. Grooming, singing in groups and religious activity all result in the release of endorphins. Thus these behaviours provide an inherent feel-good factor and facilitate social bonding. He speculates that there is an important relationship between the religious experience and group singing; hence the importance of music in ritual practices. Right hemisphere activities are the source of numinous feelings and there is an integral relationship between the holistic production of right hemisphere organisation of music and sensations which may be equated with animistic awareness.

Gell (1998) suggested that traditional societies relate to art objects as people. This is of course considering people not as bounded biological organisms but as 'all the objects and/or events in the milieu from which agency or personhood can be abducted' (222). Similarly, research suggests that the action of music on the brain is also like the action of another person (Watt and Ash). This relationship with right hemisphere activity may suggest that art and music are a means to embody the numinous experiences of the right hemisphere which led to the beginnings of religion. Gell stated that ancestral 'shrines, tombs, memorials, ossuaries, sacred sites; all have to do with extensions of personhood beyond the confines of biological life' (1998: 223). These extensions may be embodied as music and art and they may be related to persons, ancestor or spirits. We may associate music with religious and ceremonial activity precisely because it brings the numinous into the here and now (Tuzin).

Tuzin suggests that certain sound stimuli are able to produce numinous sensations and that 'religious culture stands ready, so to speak, to provide an interpretational object' (579). In the same vein, Mithen proposes that with 'the emergence of religious belief, music became the principal means of communicating with the gods' (266). Yet I would stress that the psychological and cultural context is crucial for this communication to take place. Music 'has no effect in the body or consequence for social action, unless its sounds and circumstances can be related to a coherent set of ideas about self and other bodily feeling' (Blacking 1995: 176; see also Rouget).

The position I take here views music as a method presenting concepts of order and organisation which increases the stability of the self and the society by maintaining relationships, including those with the spirit world. Some examples from ethnography will make the idea seem a little more concrete.

Ethnography and palaeo-organology
Iain Morley (2003; 2006) compared the music of four groups of hunter-gatherers (the Blackfoot and the Sioux, from the Great Plains; the Aka and Mbuti pygmies; the Pintupi of the western Australian desert; and the Inuit). All the groups meet during the difficult subsistence seasons which see an increase in ceremonial and communal music, often for pleasure: this is accompanied by rhythmic dancing. In all cases the music is predominantly vocal with some percussion but little use of melodic instruments, which when employed are crafted from natural materials. All four peoples believe that they sprang from the land—that is, their creation myths link them directly with the land in which they live. They believe that they are related to the local animals and able to influence their environment and the spirits through their music. So Morley's ethnography supports our proposed model.

These musics share three attributes: they promote group cohesion, alter mood, teach dance and support group interaction; they have no inherent symbolism but accompany symbolic activity; and at least for some groups (namely the Pintupi and Inuit), they act as a mnemonic aid. Summarising further, the key features of Morley's study which are important for us are that the music uses non-instrumental, vocal melody; employs percussion through stamping, clapping and slapping; uses wood, but rarely bone, for scrapers, flutes, drums and drumsticks; and shows little alteration of objects (for example, flutes do not need to have finger holes). This means that there is likely to be little surviving evidence. Morley is of course looking at hunter-gatherers' use of music, so exact parallels with farming societies should not be expected. But his study should give us ideas about the significance and role of music for communal interaction and about which materials may have been used.

In Britain, few prehistoric musical instruments have survived. But there are some important examples, and if we cast our eye further afield we have a great deal of contemporary evidence from continental Europe. There is good evidence for contacts, trade and the movement of people, such as the Amesbury Archer, and the range of instruments may have been similar even if they do not survive.

Flutes in Britain and France
The earliest flute in Britain was found in the cist with the primary cremation from the bowl barrow known as Wilsford G23, on Normanton down. This incomplete instrument is made from the ulna of a crane and survives to 18.7 cm in length, but because of the damage we do not know how the sound was produced (Clarke *et al*, fig. 4.52; Hoare: i, 199; Megaw 1960, pl. 2.6; Wiltshire Heritage). This is a consideration of interpretation for others periods as well. Did an instrument have a duct, like a recorder? Was it played in the style of a classical flute? Was it played like an end-blown flute such as a ney? Or a tongue duct flute? Did it have a reed inserted in the end? Working models made from goose bone may give an idea of the quality of sound, and may be played as a diagonal flute or with a small reed inserted in the end.

Another roughly contemporary flute comes from the barrow known as Avebury G35. On 19 July 1849 John Merewether, Dean of Hereford, pottered out to identify some nice burial mounds to dig in, about a mile and a quarter from Avebury. The following day he uncovered this one, which revealed the crouched inhumation of a man, a coarse undecorated urn containing the bones of a child, and a tube of bone with three holes (Fig. 3.1 b). The whereabouts of this likely flute are currently unknown.

The contender for the oldest dated bone flute from Britain comes from Penywyrlod, with a radiocarbon date range of 3960–3640 BC (Darvill 2004b: 136, 256). It consists of a fragmentary sheep femur, with the remains of three holes; it may be unfinished. The closeness of two of the finger holes, 15mm from centre to centre, has been cited as evidence against a musical interpretation. But we should not judge all musical intervals by our own standard: while western music divides an octave into 12 notes, in India the octave may have 22 divisions, although only seven of these notes will be used. We should not expect modern pitch differences in the past.

Despite resembling a flute or pipe, the lack of a voicing lip has also been noted to call into doubt this interpretation (Darvill 2004, fig. 55G; Megaw 1984), although Megaw suggests that the partial hole may be a notched lip. Yet the need for a fine lip to produce sound may not be as important as often argued (Wyatt 2008; Montagu, pers comm). Models, demonstrated by the author, with a roughly cut edge, a smooth curved edge and a fine lip all produce essentially the same tone. If may be that this object is the result of animal gnawing (Helen Leaf, pers comm). Another possible flute comes from Skara Brae, with a date between 3200 and 2600 BC, but like the previous example it is very roughly worked and its interpretation as an instrument is disputed (Brundel). A final example, a crane tibia with four holes, comes from near Lincoln, but it may be of far later date (Megaw 1960).

So far we have looked at the fragmentary British examples, but a complete vulture bone flute was discovered in a cave used for burials at Veyreau, Aveyron (Fig. 3.1 a). It was sealed in a limestone concretion with human bones which have been dated by radiocarbon to 3800 ±130 BP (approximately 2500–200 BC). The flute has five circular finger holes and a small square window at one end. Another small hole near the end furthest from the mouth may have been intended to correct tuning. It may have had a block of wax or wood and have worked in the manner of a duct flute, while blocking the mouth end with the bottom lip allows it to be played in the style of a notched or transverse flute which has been rotated 90 degrees. Each of the flutes mentioned above was found in a funerary context, which might suggest the use in a funerary ceremony or a personal item accompanying the dead to the afterlife. Photographs of the Veyreau flute may be found in Fages and Mourer-Chauviré (figs. 3–5) and Scothern (pl. 36).

Whistles, wooden pipes, natural flutes and a globular clay flute
A simpler form of these flutes may be seen in phalange whistles. These are known from the Palaeolithic, generally made from the toe bones of deer. Yet in the Neolithic of Britain they are made of cattle toes, and we may note that the deposition of cattle remains is part of a significant ceremonial tradition (Burl; Midgley; Pollard and Reynolds). Indeed, some cattle skulls found near the third entrance through the Stonehenge bank have a date several hundred years earlier than the commencement of the building project. At West Kennet long barrow, six of these perforated cattle toes are known (Fig. 3.1 e; colour photographs found in Clarke *et al*, fig. 2.8, and at Wiltshire Heritage). Similar artefacts are rare but are known from the Lower Dounreay chambered tomb, the earlier levels at Jarlshof, and Skara Brae (Clarke *et al*, fig. 5.63). These pierced bovine phalanges are still found in the Bronze Age, as at the site of Le Fort-Harrouard, Eure-et-Loire (Clodoré 2002a, fig. 23). We may see a variant of this during the Bronze Age burials in a class of object commonly named toggles. When these objects have a single hole in the side they may equally be used in the manner of a phalange whistle. The example from the Sewell burial is 2 cm in length (Fig. 3.1 c), but a decorated example from Westerdale, Yorkshire, is 4 cm long with a central hole (Clarke *et al*, fig. 7.30). We may compare this with the Köping, Scania, find (Fig. 3.1 f) which Lund (1991) compares with recent flutes used by hunters to mimic otter calls.

Instruments made from plants are unlikely to be preserved except in special circumstances, but a few examples have survived. From the lake settlement at Charavines, Isère, a 42 cm tube made from hollowed elder has survived (close-up of end illustrated in Fig. 3.1 d; see also Clodoré 2002a, figs. 16a and b). It has no finger holes but may still be used as a flute: we may remember Morley's observation that flutes often have no finger holes in hunter-gatherer societies. A tube of these dimensions may play eight distinct notes based on the principle of blowing at the end as with a ney and increasing the pressure of breath while alternately stopping and unstopping the end. Yet while it is possible to play sounds with this object, it may simply be a tube designed to fan the flames of a fire, as was used until recently in the Lot region of France (Clodoré 2002a); such items are still sold in some tourist shops. Bognár-Kutzián records a tube not from wood but red deer, 18 cm in length and with an internal diameter of 2 cm, found in grave 52 at the Tiszapolgár-Basatanya cemetery; like the Charavines object, this piece of hollowed antler has no finger holes (pl. 59; compare Lund 1991, fig. 9).

In 2003, archaeologists excavated a set of six wooden pipes from a cooking site related to a Bronze Age settlement of Charlesland, south of Dublin. Six tubes of yew were found in a wood-lined trough. The longest tube is 50 cm, and a radiocarbon date from the trough gave a date range of 2137–1909 BC. There are no finger holes in the tubes and no parallels are known, but two different interpretations have arisen, suggesting that it was a simple organ with the pipes with detachable ducts or reeds fitted in a frame, or was played like a large pan pipe (Gowen; O'Dwyer 2004, pl. 23).

In Hungary, Greece, Italy and France, conch shells with the end removed may have been used as trumpets (Clodoré 2002a; Montagu; Skeates). No conch shells are known from the Neolithic or Bronze Age of Britain. What is interesting in the case of the Italian conch shells is that they are often associated with burials in caves. However, one may achieve surprising tones with a small whelk or garden snail. Hollow flints may be used in a similar fashion, three examples found in Denmark (two from settlement contexts at Grund, Fausing, the third from a field at Viderupkjaer, Farstrup) have been interpreted as whistles. Two of these objects had the embouchure worked to make it easier to play and one of them has been given a single finger hole (Lund 1991). Along similar lines, a whistle manufactured from a naturally hollowed geode was discovered in the settlement site of Fort-Harrouard, dating to between 4200 and 3500 BC (Clodoré 2002a), and examples are known from Scotland and Ireland (Purser).

From a Neolithic settlement at Mramor, north of Veles in Macedonia, comes a clay globular flute. The flute has no context, having been discovered in a ploughed field, but the dates of the settlement range from 5000 to 4000 BC. The complete instrument is made of fine reddish clay with some darker areas due to burning; it is undecorated. It measures 4.7 cm in diameter and has three holes, two 0.4 cm and one 0.6 cm wide, forming a triangular pattern (images accessible via Jovcevska). This is a little early for us but gives a demonstration of the type of technology available in the Neolithic of Britain. Another Macedonian example dating to around 3000 BC comes from Grlo, but this is a simple duct whistle with no finger holes. Using a working model based on the Mramor flute, albeit with a slightly larger diameter, it is possible to achieve three notes as a standard and additional changes by half covering the holes and changing the pressure of the breath.

Horn, drums, bullroarers, musical bows and rattles
By the later Bronze Age, two styles of elaborately manufactured bronze horn are known in Europe: the Irish horn, thought to have been played in the style of a didgeridoo, and the lur, found in the lands surrounding the Baltic sea, with a trumpet-style mouthpiece. Both styles of instrument had been deposited as pairs (O'Dwyer 2004, pl. 4; Lund 1991: 51). The lurs are known to anyone who has looked closely at the wrapping of Danish butter. Their elaborate manufacturing suggests these horns had a long history, no doubt originating in the use of animal horns. Two clay examples were discovered in rock shelters in southern France, dating to

the final Chalcolithic around 2500 BC. The trumpet of Vallabrix measures 35.5 cm, with a maximum diameter of 9 cm. The example from Rouet, Hérault, is 32 cm long (Clodoré 2002a, figs. 19–20). Again we have no similar examples from Britain, but it is entirely plausible that animal horns or rolled bark instruments were used and have not survived; these are still found in various parts of Scandinavia, and were played in Britain as whithorns during the early 19th century.

While we have no clear drums from British contexts, our continental neighbours had a profusion of them. They stand out on the continent because they are made from clay, and it is likely that wooden examples would have been known outside the area of central Germany, Denmark and Poland, where they are found (Fig. 3.2). There is some evidence that they were also found in the Neolithic of Hungary and the Ukraine (Behrens 1979; Wyatt 2007). The examples from Germany are associated with remains that archaeologists call the Funnel beaker culture (TRB), and date to between 3350 and 2700 BC. They are found in settlements, in burials accompanying men, women and children, and occasionally in ceremonial sites. They are commonly fragmentary, although complete examples have survived. The largest example is 46 cm high, the smallest 4.5 cm.

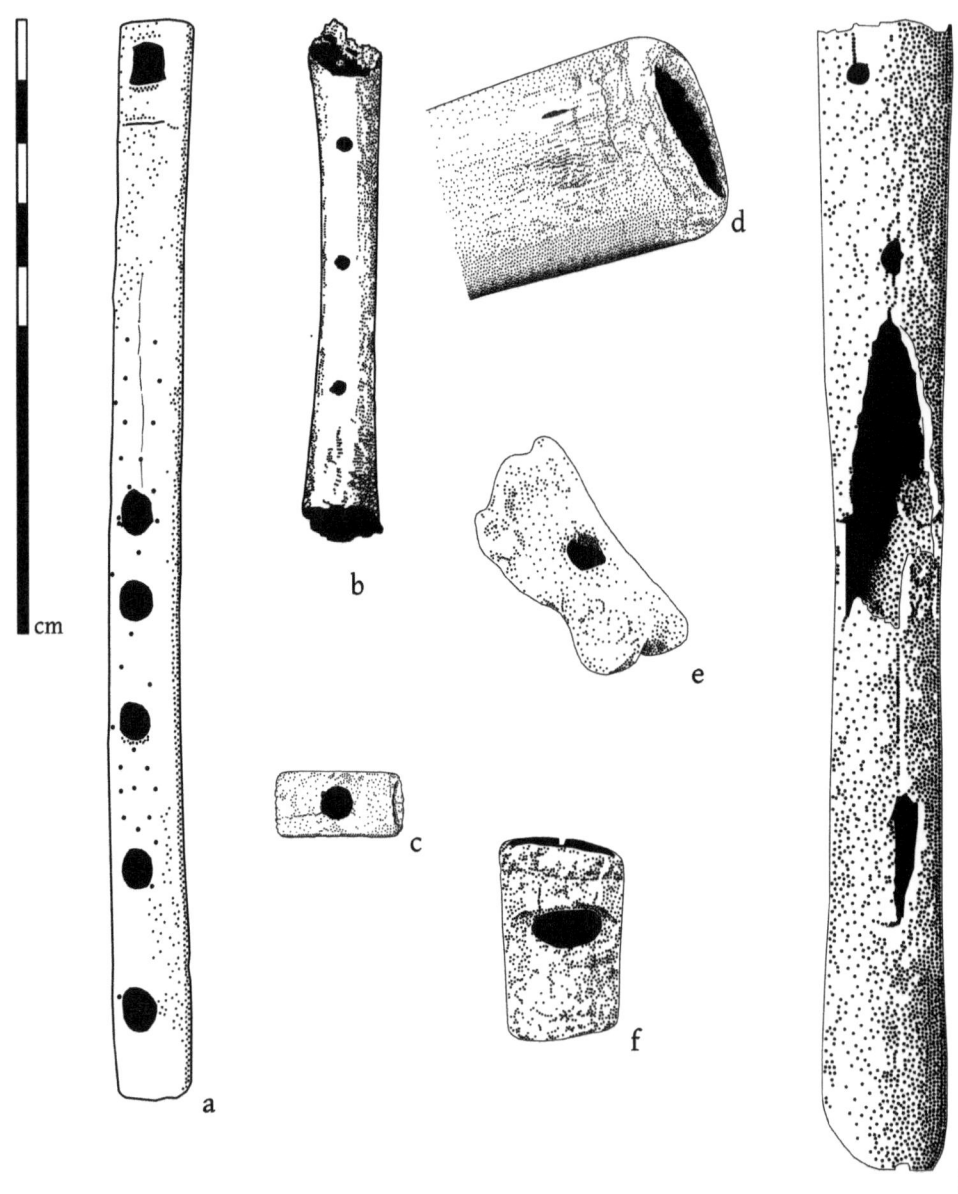

FIGURE 3.1: PREHISTORIC FLUTES AND WHISTLES FROM VEYREAU (a), AVEBURY (b), SEWELL (c), CHARAVINES (d), WEST KENNET (e), FALKÖPING (f) AND TISZAPOLGÁR-BASATANYA GRAVE 67 (g). IMAGES DRAWN FROM PHOTOGRAPHS IN BOGNÁR-KUTZIÁN, CLODORÉ 2002A, LUND 1991 AND PIGGOTT 1962; AFTER A DRAWING IN MEREWETHER; AND AFTER A PHOTOGRAPH BY N WYATT.

The sherds of the Böhlen drum, from the Harth forest (Fig. 3.2 a), were found in a burial mound measuring 20 x 30 metres. The drum had been broken and placed beneath a paved area consisting of the settlement sherds on which lay the body of a man (aged 50–60), facing east in a crouched position. He had twice survived trepanation. The drum was 22.6 cm high with four vertically pierced lugs, placed below the rim. The decoration of this instrument was confined to the foot, both inside and out (Behrens and Schröter; Fischer; Mildenberger 1953; J Müller; Wyatt 2007, 2008c).

These drums are hollow in the style of a *jembe* or *darabukka*, so they cannot have been pots, though a skin may also have been attached to ordinary clay bowls for them to function as drums. They have distinct forms, allowing us to distinguish between drums associated with settlement and the living and those associated with burials and the dead. Similarly, the decoration on the TRB drums is not found on other objects of the TRB societies and shows a distinction between burial and settlement decoration (Wyatt 2007). The decorative motifs bear a close similarity to entoptic phenomena, also known as subjective visual phenomena (Dronfield 1995, 1996). These are geometric shapes produced in the optical system during the altering of consciousness, and this may be related to the fact that drums are capable of inducing changes in brainwave frequencies through entrainment (Maxfield). This would make the drums ideally equipped to accompany funerary ceremonies, since a key role of shamanistic practitioners is to play the part of the psychopomp who accompanies the souls of the deceased to the land of the dead (Price-Williams and Hughes; Vitebsky; Wyatt 2010a and b).

Three further examples of instruments available to our ancestors conclude this overview. First, bullroarers—oval pieces of bone or wood—are known from the Palaeolithic. The Kongemose bullroarer from Denmark dates to 6500 BC (Lund 1991). Native American coming-of-age rites used up to sixty bullroarers at the same time and they were still given to boys as a symbol of manhood in Sweden in the middle of the 20th century (Lund, pers comm). Second, the use of a bow as an instrument may have begun with the testing of the tension of the string (Lund 1991). An example of a bow found in Ageröd, Sweden, dates to 6000 BC. Being 75 cm in length, it unlikely to have been a hunting bow; it may of course have been part of a drill set or used in fire making (Lund 1991). We should also note that in Greece and the near east there is evidence for multiple stringed instruments in the 3rd millennium BC. Finally, rattles may have been made from woven grasses or pliable wood such as willow, and as such we do not know the age of this style of instrument, but once they are manufactured in clay it is a different story. The earliest example from Europe comes from Roßleben, Thuringia, and was discovered in close proximity to five burials of Linearbandkeramik date (LBK), that is, the late 5th to early 4th millennium BC. It resembles a pedestalled bowl, but the foot of the vessel is hollow and contains small pieces of clay to produce the rattle (Kaufmann). Another example of a clay rattle, from Vykhvatintsi, dates to the 4th millennium and was associated with the late Cucuteni culture (Gimbutas, fig. 154).

The more one experiments with methods of making sound, the more it is clear that most objects, be they an instrument or not, are capable of being used as such. This does not mean they all were instruments, but that we should be neither over-optimistic nor over-pessimistic in how we interpret artefacts and structures (see d'Errico and Lawson).

Crescendo: spirit music and ancestral flutes

I began by suggesting that music was linked to activity intended to stabilise the individual and the community. Part of this involved its ability to allow people to communicate with spirits and the ancestors and be able to affect their environment through the action of those beings. I have noted that while the archaeological record is not rife with instruments in Britain, it is clear that the millennium leading up to the building of Stonehenge was marked by its special treatment of the ancestral bones. Removal of bones may be recognised archaeologically, and in Britain this was found to have been done between 4200 and 3000 BC. At West Kennet there were fewer skulls than internments, and a large number of these may be represented at the Windmill Hill causewayed enclosure, visible from the site of West Kennet itself, maybe for reasons of denying the individual and emphasising the collective or as a way of keeping the ancestor within the community for longer—a kind of extension of personhood (Parker Pearson, 1999: 52; see also Gell 1998). Here we might just be allowed to recognise an overlap with the earliest phase of Stonehenge. But cattle skull, jaws and a deer tibia from a possible third entrance at Stonehenge have been dated to 3400–3200 BC, suggesting at least in the case of animal remains that some curation of bone was taking place at Stonehenge itself (Burl; Pollex)—a further link between Stonehenge and the preceding traditions. West Kennet itself has dates ranging from 3783–3373 BC cal in the northwest chamber to 3633–3104 BC cal in the northeast chamber (Pitts 2001: 279). Further afield, from France to Hungary, this was also the case, but with a slight note of difference.

I have come across four examples of flutes made from human femurs or ulnas. In Italy, at Val Rossandra and Trentino, flutes were discovered made from femurs (Megaw 1960; Grant *et al*, fig. 8.19). The Trentino example was elaborately decorated. At the cave of Las Morts, in France, a flute was made from a human ulna (Clodoré 2002a; Fages and Mourer-Chauviré, fig. 6.24; Gailli: 105; Scothern, fig. 2). Finally, at the Neolithic cemetery of Tiszapolgár-Basatanya, Hungary, a human femur in grave 67 with multiple perforations accompanied the burial of an old man (Fig. 3.1 g). This maybe another example of an ancestral flute. In the grave next to this individual had been deposited a vessel which may well be a drum (Behrens 1979; Wyatt 2007).

These human examples are strictly speaking earlier than the visible structures at Stonehenge, but the dates are continually creeping back. It is also clear that sites such as Stonehenge were important locations much earlier than the visible remains may suggest. A series of four huge posts just outside Stonehenge have provided a radiocarbon date between 8500 and 7650 BC (Pryor). This is an important continuity of ceremonial space. In this context maybe a re-examination of some of the human bone remains in Britain may find signs of similar practice and represent an attempt to use music to affect the spirits using instruments made with those spirits' own remains.

I would like to thank my father and Chris Scarre for commenting on previous drafts of this chapter.

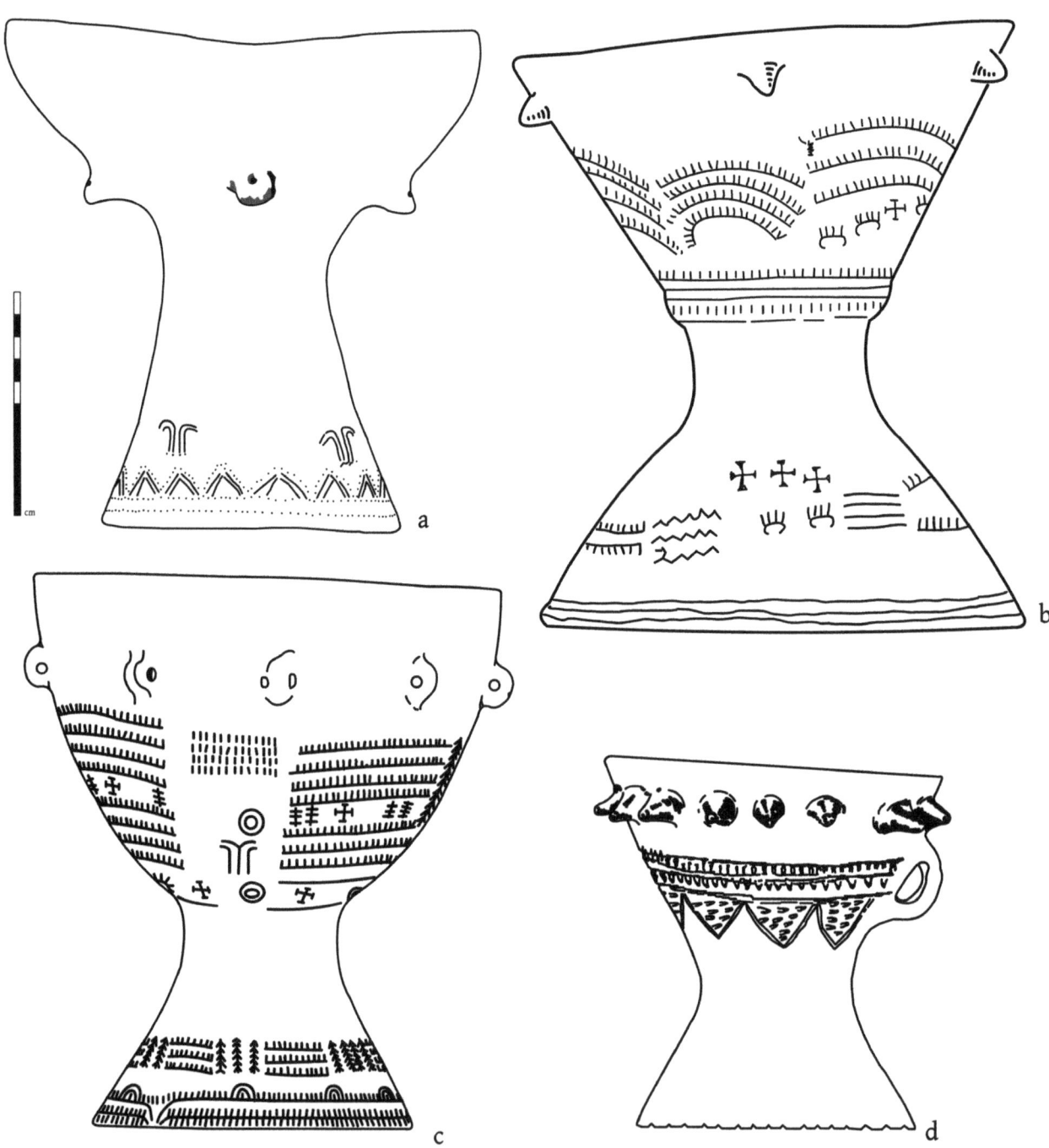

FIGURE 3.2: PREHISTORIC DRUMS FROM BÖHLEN (a), ZORBAU-GERSTEWITZ (b), HORNSÖMMERN (c) AND LANGENBURG PIT 95 (d). IMAGES REDRAWN AFTER BEHRENS AND SCHRÖTER, MILDENBERGER 1952, D W MÜLLER AND NITZSCHKE.

CHAPTER 4

Songs of the stones: the acoustics of Stonehenge

Rupert Till

Introduction

What remains of ancient monuments are architectural fragments which can 'allow us to think through the orientation of the practices which both created that architecture and which were staged within it' (Barrett: 14). Acoustic analysis is a sonically based architectural analysis that can reveal detail about these practices. Because 'time is collapsed for the archaeological observer' (*ibid*: 12), even a partial or fractured understanding of the use of music, acoustics and sound in a space can act to animate the information we have from these architectural fragments, since 'sound is a sensation, and belongs to the realm of "activity" rather than "artefact". Sound brings the world to life, it can appear to fill spaces, create atmospheres, and have an intense emotive power' (Watson 2001a: 180). Existing in time, it can add a third dimension to an otherwise flattened interpretation, like pressing play on a paused DVD. While architecture demarcates space, sound demarcates time.

An archaeological study of a physical space is lifeless without an accompanying understanding of the narratives that developed within it, and the way they move through time, change, and develop. As Giddens puts it: 'A person's identity is not to be found in behaviour, nor—important though this is—in the reactions of others, but in the capacity *to keep a particular narrative going*' (54). Speaking of traditional cultures, he says: 'Where things stayed more or less the same from generation to generation on the level of the collectivity, the changed identity was clearly staked out—as when an individual moved from adolescence into adulthood' (33). Understanding these time-geographies of space, these developing narratives and rites of passage, is important so that we can understand how the users of a space felt about it. This can be explored more readily from an understanding of the use of sound in a space than by looking at the physical space, focusing on the visual.

In prehistory, sound would have been perceived differently than it is today. We are surrounded by sounds that are not part of nature, noises from machines, loudspeakers, televisions and mobile phones, ranging from the aircraft in the sky to the ipod earphones in our ears. At a time when stone buildings were rare, spaces with acoustics that were not part of nature would have been very acoustically striking to those who entered them. A man-made acoustic of any note was an extremely unusual thing.

Sound was a primary focus for accumulating knowledge, culture and information in prehistory, as a great deal of transmission happened using language (the acoustic), rather than writing (the visual). 'Among peoples at an "oral-aural" level of culture to whom writing was unknown, the ear exercised an overwhelming tyranny over the eye' (McLuhan: 28). We can therefore expect to find out as much about the reasons for the layout of a site by investigating its acoustics as by investigating its visual and physical properties. This is perhaps especially the case in a site that does not seem to be designed for strictly functional purposes such as accommodation or defence, for music and sound are often at the core of cultural, communal and ritual activities. Sound gives 'information about the temporal structure of the event that caused it and the vibratory frequency of this event . . . with great precision', and 'man likes to make sounds and to hear them' (J J Gibson: 17). In early hominid society, 'suddenly or gradually, the voice itself came to be used as a sort of tool' (*ibid*: 25), and the acoustic environment of this sonic tool, as a key focus in prehistory of communication and development within an oral and aural culture, merits as much investigation as the context of the use of bone, stone, metal, wood or ceramics.

Because it operates in the time domain, acoustics can give us invaluable information about so-called non-material or intangible elements of culture such as music, ritual and religion (J J Gibson: 26). 'When it comes to affairs of the soul, of emotion and feeling, or of the "inwardness" of life, hearing surpasses seeing as understanding goes beyond knowledge, and as faith transcends reason . . . Vision in this conception, defines the self individually in *opposition* to others; hearing defines the self socially in *relation* to others' (Ingold: 246–7). Both Gibson and Ingold discuss how 'vision and hearing are not so much disparate as interchangeable' (Ingold: 245), are an active single task of perception, as looking and listening. If we accept that 'looking and listening' is together a fused irreducible act of perception, then any analysis of architecture must include analysis of sound:

> Vision, since it is untrained by the subjective experience of light, yields a knowledge of the outside world that is rational, detached, analytical and atomistic. Hearing, on the other hand, since it rests on the immediate experience of sound, is said to draw the world into the perceiver, yielding a kind of knowledge that is intuitive, engaged, synthetic and holistic. (*ibid.*)

Archaeoacoustics is a fairly new field of interdisciplinary study that attempts to uncover information about ancient cultures from a study of the acoustics of specific sites. Work related to this field includes a study group focusing on music archaeology (the International Study Group on Music Archaeology) which has developed from work within the International Council for Traditional Music, and a recent book from the McDonald Institute, Chris Scarre and Graeme Lawson's *Archaeoacoustics* (2006), the first collection of writings specifically on this subject. In *Archaeoacoustics*, and in another paper (Watson and Keating 1999), Aaron Watson discusses his pilot study of the acoustics of the existing Stonehenge site. He draws attention to the interesting acoustic features he found, and to possible evidence that they were part of intentional acoustic design. In response, I carried out my own theoretical acoustic consideration of the final phase of the complete site in prehistory, which made it clear that powerful acoustic effects within the site would have been likely and that there was a substantial amount of further work to be done on the subject.

The project described in the remainder of this chapter was begun in July 2007 in order to investigate further the acoustics of Stonehenge. The work was undertaken by myself along with Dr Bruno Fazenda, an acoustics specialist. It asked what could be discovered from the study of the acoustics of Stonehenge about life in prehistoric Britain at the site and in general. It aimed to investigate the extent to which music and sound were a part of ritual activities at the site, and to uncover information about the nature of such activities. It attempted to illustrate what could be discovered about an archaeological site from an in-depth consideration of acoustics and sound, even where that site had already been extensively investigated archaeologically.

Research context
It was important when investigating such a well-studied and prominent monument to operate within the context of existing archaeological knowledge. This project aimed from the beginning to work with archaeologists who had detailed knowledge of the site. Professor Mike Parker Pearson and members of his Stonehenge Riverside Project, a multi-university large-scale ongoing archaeological team investigating Stonehenge and its surrounding landscape, provided advice on analysing the acoustics of Stonehenge. Parker Pearson's theories suggest that Stonehenge may have acted as a place of the dead rather than the living, and as a centre of funerary and ritual activity. The Stonehenge Riverside Project contextualises the stone circle within its landscape, including the Durrington Walls site and others, and it was possible to discuss ideas, theories, principles and various details with members of the team from time to time, as well as to visit their onsite archaeological excavations in August 2008. When results are completed, it is intended that this archaeoacoustics project will provide information for the ongoing research by the Stonehenge Riverside Project and other archaeologists interested in the site. It is not intended to provide definitive solutions or insights but to be a part of the archaeological jigsaw of the site, to contribute a small part to our understanding of Stonehenge and its landscape. It is not thought that sound and acoustics alone will lead to some kind of unlocking of the secrets of the site, but that they will contribute to our growing understanding of it.

There are many theories about and issues relating to the contemporary, historic and prehistoric uses of Stonehenge. Many of them relate to ritual use of various kinds, which would be likely to involve music or at least use sound. Nettl describes the association of music and ritual 'in addressing the supernatural' as a universal, 'shared by all known societies, however different the sound'. He continues: 'Another universal is the use of music to provide some kind of fundamental change in an individual's consciousness, or in the ambience of a gathering. . . . And it is virtually universally associated with dance; not all music is danced, but there is hardly any dance that is not in some sense accompanied by music' (469). Ehrenreich agrees: 'These ingredients of ecstatic rituals and festivities—music, dancing, eating, drinking or indulging in other mind-altering drugs, costuming and/or various forms of self-decoration, such as face and body painting—seem to be universal' (18). Traditional musical forms are rarely set apart from the social and cultural context. In fact in numerous traditional cultures there is no separate word for music, with terms instead meaning music and dance, or music and trance:

> In the African world music has played such a central role in the life of its people for so long that there is often no separate word for it in indigenous languages. Like religion, music permeates the societies of sub-Saharan Africa in a way difficult to understand in the west. An essential vehicle for communicating with God and the ancestors, a key determinant in rites of passage from birth to death, a tool for healing the ill, educating the young, settling disputes and entertaining the communities of both rural and urban Africa, music is perhaps *the* essential foodstuff for the African mind, body and spirit. (Gray: 15.)

While it is not suggested that Stonehenge necessarily featured music and dance exactly like that in traditional African culture, it is likely that music was integrated into participatory ritual that included movement and/or dance.

Recent studies of the origins of music include two books, *The Singing Neanderthals* (Mithen) and *The Origins of Music* (Wallin *et al*). In the latter, Walter Freeman makes it clear that cultural music-related activities are hard-wired into humanity through biological and evolutionary development:

> Music and dance originated through the biological evolution of brain chemistry, which interacted with the cultural evolution of behaviour. This led to the development of chemical and behavioural technology for inducing altered states of consciousness. The role of trance states was particularly important for breaking down pre-existing habits and beliefs. That meltdown appears to be necessary for personality changes leading to the formation of social groups by cooperative action leading to trust. Bonding is not simply a release of a neurochemical in an altered state. It is the social action of dancing and singing together that induces new forms of behaviour, owing to the malleability that can come due to the altered state. It is reasonable to suppose that musical skills played a major role early in evolution of human intellect, because they made possible formation of human societies as a prerequisite for the transmission of acquired knowledge across generations (422).

The practice of musical skills continues to serve a similar role today. Many writers make it clear that music and dance, often as part of ritual or religious activities, play a key role in building and maintaining community.

Ehrenreich states that 'rituals serve to break down the sufferer's sense of isolation and reconnect him or her with the human community . . . because they encourage the experience of *self-loss*, that is, a release, however temporary, from the prison of the self, or at least from the anxious business of evaluating how one stands in the group or in the eyes of an ever-critical God' (152). Musical activities in rituals such as may have occurred at Stonehenge are part of what Durkheim terms 'collective effervescence' and Victor Turner *communitas* (Ehrenreich: 14; Turner); they are technologies for the building of communities. Events at Stonehenge in recent history, even if presented as secular, are still ritual activities, what might be described as implicit rather than explicit religion (Bailey).

It may be that music and sound could give us insight into how competing interpretations of Stonehenge can be related to one another. Darvill's recent theory, developed following a 2008 excavation within the stone circle itself, describes Stonehenge as a place of healing, something that may seem very different from Parker Pearson's theories of Stonehenge as a place of the dead. However, communicating with the dead or the spiritual world in rituals such as those that feature music-based trance practise can be a healing activity as well as relating to funerary rituals, the ancestral spirits from another world coming to communicate and provide physical or spiritual healing. Thus discoveries about music and sound at Stonehenge may help to resolve and integrate theories that currently compete and contrast, a fitting analogy for the communal and integrative power of music for the healing of social or emotional health. Considering music as a form of communal or social technology, it should be studied in as much detail and with as much rigour as any other feature of the site.

So we can perhaps predict that at Stonehenge this project may find music part of a ritual activity and addressing the supernatural; integrated into its social context; combined with dance; designed to lead to the achievement of altered states of consciousness or trance or to change the ambience of a gathering. It may find it used both to bond the community and establish one's position within the community; may find it to involve communication with gods and the ancestors; to be part of rites of passage from birth to death; to be used as part of healing of some sort.

Although we can strongly suggest that there would probably have been music at Stonehenge, it is difficult to define what that music would be like. There is no historical record of British prehistoric music, and there is little information in the archaeological record. We have some ideas of the kind of instruments that might have been played, as can be seen from the work of Simon Wyatt elsewhere in this volume. Aaron Watson's recent work, also described in this volume, has involved creating multimedia artworks that aim to give a phenomenological impression of the soundworld of prehistoric Stonehenge. The project described in the present chapter has taken another approach, aiming instead to try to find in the acoustics of Stonehenge echoes of the music that was originally played there, evidence that might help us to describe this music.

One problem with existing archaeoacoustic methodologies is the nature of what currently exists at Stonehenge. Even if one sets aside the considerable reconstruction carried out at Stonehenge during the 19th and 20th centuries, the site is a collapsed remnant of the one that existed in prehistory. One approach to studying the acoustics of Stonehenge is to undertake acoustic field measurements of the current site. Useful though this is, it does not tell whole story. Alone, it is akin to standing within a ruined abbey and hoping to define the acoustics of an intact cathedral.

Watson's published pilot study was based on such field tests. It aimed to make the best possible use of the site's most intact section by being focused on the different acoustics inside and outside the outer sarsen stone circle. (Stonehenge 3vi, as the final phase of the monument is generally described, has some stones made of sarsen, a type of sandstone. In particular, the outer circle and its ring of lintels are made of this material.) The use of an acoustics-based field test approach raises further problems. Such methodologies are designed for indoor enclosed spaces, often to address health and safety issues for existing buildings or to enable architects to consider acoustics in their designs. Acoustic consultants do not generally use their techniques to forensically examine the acoustics of part-demolished sites or buildings. True, existing techniques and methods are often designed to identify unwanted acoustic effects, but they do not allow such effects to be studied in detail, rather than merely identified and removed. Techniques from acoustics as applied to buildings are of course useful, but this project aimed to establish a new broad approach involving cutting-edge technology.

Methodology

The first stage of the project was a literature review. This was followed by a mathematical acoustic study of archaeological plans of Stonehenge. A further stage of the project was the acoustic study of a physically constructed model of Stonehenge. The final stage was digital acoustic modelling, the use of computer software to analyse the acoustics of a digital, graphic model of Stonehenge. From comparison of these theoretical studies and the results of a literature review, a hypothesis could then be developed, duly tested by looking for evidence to support it from onsite field measurements and tests at Stonehenge itself.

The final stage would be the reconstruction of the acoustics of a site, conducting experimental archaeology using sounds. It would be important to note that any reconstruction could not be a definitive explanation of the acoustic properties of Stonehenge, but that it could be a useful process, helping to give a phenomenological understanding of the site and providing an indication of the likely or possible sound, acoustics and music of prehistoric Stonehenge.

With very limited funding for this work, the project aimed to conduct an initial study into each area. Further work would include construction of virtual realisations of Stonehenge with integrated imagery, sound and acoustics. It would also be important to make comparisons with other sites and stone circles, and with other parts of the surrounding landscape as well as the Stonehenge stone circle. Although work is still continuing on the project, it is appropriate to present an indication of initial results.

Literature review

The literature review included an investigation of archaeological publications about Stonehenge. It also included a review of publications on the origins of music, music in prehistory and archaeoacoustics. Key texts included those already cited in this chapter, plus Cleal *et al*, Devereux 2001, Reznikoff and Dauvois, and Rouget. There exists however a wide range of other relevant publications. A current Science and Heritage Research Cluster (Acoustics and Music of British Prehistory Research Network), focusing on the acoustics and music of British prehistory and funded by the AHRC and EPSRC, has created an online reading list.

Within English literature, Thomas Hardy, living in Wessex and writing about it, provides perhaps the best known historical evidence for acoustic features at Stonehenge. He refers to the monument in a number of ways. In his autobiography, he mentions having revisited the monument in 1897, and Millgate discusses Hardy visiting it a third time in 1899 (F E Hardy; T Hardy *et al* 1984: 400–401). Hardy also appears to have had two sarsen stones in his garden at Max Gate, Dorchester, and his works contain a number of 'pagan' references (Cope; Harrington).

In *The Trumpet-Major* Hardy refers to 'the night wind blowing through Simon Burden's few teeth as through the ruins of Stonehenge' (217). Here, in 1880, Hardy is discussing the sound made by wind rushing through the gaps between Stonehenge's sarsen uprights. He makes more detailed reference to echoes and resonances within Stonehenge in his 1891 novel *Tess of the D'Urbervilles*. In it he writes:

> All around was open loneliness and black solitude, over which a stiff breeze blew . . .
> 'What monstrous place is this?' said Angel.
> 'It hums,' said she. 'Hearken!'
> He listened. The wind, playing upon the edifice, produced a booming tune, like the note of some gigantic one-stringed harp. No other sound came from it . . . At an indefinite height overhead something made the black sky blacker, which had the semblance of a vast architrave uniting the pillars horizontally. They carefully entered beneath and between; the surfaces echoed their soft rustle; but they seemed to be still out of doors . . .
> 'What can it be? . . . A very Temple of the Winds' . . .
> 'It seems as if there were no folk in the world but we two' . . . they . . . listened a long time to the wind among the pillars . . . Presently the wind died out, and the quivering little pools in the cup-like hollows of the stones lay still (501–4).

Hardy describes the sense of solitude at the place, the sense that the outside was shut out, a sense of envelopment and enclosure. He says that the wind was blowing strongly and made the site hum, producing a 'booming tune'. This could be evidence of low frequency resonance, standing waves set into sympathetic vibration by the wind; this could also explain the vibrations in the pools of water in the hollows of the stones. Without the motor traffic of modern times, this kind of resonance may have been more readily audible in the 19th century. Indeed the road noise may now mask such a sound, and a low hum on windy days may be missed or mistaken for passing traffic. It seems very likely that Hardy had visited the site, had heard this sound, and decided to use it in his work.

Hardy also references echoes made when first entering the space, at the edge of the circle, another interesting acoustic feature. He implies that the space seems at first, on entry, as though it were an indoor stone-built space, with echoes, but that one was actually outdoors. Tess, the lead female character, lies on a central 'altar' (probably the 'Altar Stone', the place where the audio effects would have been most prominent), listening to the wind in the pillars.

Hardy refers to the humming sound as like the note of a one-stringed harp. He is referring to the sound of an Aeolian harp. This is a wooden box with metal strings, which is placed so that the wind blows on the strings, and produces varying pitches. It is known that such instruments were made in ancient Greece, and they also became popular household items in the romantic era. Coleridge wrote a poem entitled *The Aeolian Harp*, and Hardy has one of his characters build one in *The Trumpet-Major*. Indeed, he was especially interested in the sonic: in music (he was a violin player), in the sounds made by nature in general, and in sounds made by the interaction of the wind with nature in particular. Prior to *Tess of the D'Urbervilles*, he had described the wind activating the trumpets and bells of flowers on the heath in *The Return of the Native* (40), and the different sounds of the wind in various trees in *The Woodlanders* (64). Michael Irwin comments:

> The noises in Hardy's novels, like the visual descriptions, are there to remind us that human beings are surrounded by contingent phenomena, overwhelming in scale and diversity, which are sometimes an influence on us, sometimes a distraction, always a source of information and analogy. More specifically . . . sound stands for something beyond itself. It is produced by action, something happening, something being done. It affirms that we live in a world of endless, restless movement (68–94).

In one of his chapters, Irwin analyses the use of natural sound within Hardy's novels as cinematic, used to bring the text alive in the present.

Most significant of all is information from Harold Orel, who reproduces an interview with Hardy from the *Daily Chronicle*, 24 August 1899: 3, on the subject of the possible sale of Stonehenge. We are told by Hardy, 'I have no more knowledge of the monument than is common to, or obtainable by, anybody who chooses to visit it'. Speaking of research to write *Tess*, however, Hardy says he 'made special visits to Stonehenge to get his lights for the chapter' and that he 'lives within a bicycle ride of it'. He explains the low sound he describes in *Tess* by saying that 'if a gale of wind is blowing, the strange musical hum emitted by Stonehenge can never be forgotten' (196–200). This would seem therefore to have been a well-known feature of the site at the time.

To summarise, Hardy notices a sense of enclosure and envelopment in the space, a sense of an indoor stone acoustic outside, echoes when entering the sarsen circle, interesting sound at the centre, and a low-frequency booming hum that was caused by the wind and evidenced by vibrations in water. It would be interesting to see if this project could confirm any of these observations.

Acoustic analysis from archaeological plans

The project was to be focused principally on phase 3vi of Stonehenge's stone circle, with the hope of future work on other phases. Using archaeological drawings of Stonehenge, a basic acoustic analysis of the site was carried out. This was done by considering reflections of sound from various sources, calculating delay times of principal reflections, and assessing likely reverberation and likely positions and modes of standing waves. It was immediately apparent that the sarsen stone circle, and in particular its circle of stone lintels, would be a highly significant acoustic feature. Standing just inside the circle would be acoustically significant, as edge effects by the wall would boost low frequencies.

Ray paths were traced for the transmission of sound in the space, taking into consideration materials, surfaces and reflections. It was clear that sound would move most freely between the outer sarsen and the bluestone circles. (The earliest stones present, and the principal other type of stone at the site, are often referred to as bluestones. These are mostly made from spotted dolerite from Preseli in Wales. Recent work by Paul Devereux has shown that a village in the area was called Maenclochog—Ringing Stones—and that it was known that stone in the area had particular acoustic effects [Devereux and Wozencroft].) It seemed possible from this analysis that there might have been some sort of partial whispering-gallery effect. This effect allows the whispered speech of someone standing distant from a second person to be heard clearly, even though the distance would normally mean nothing could be heard. It is often observed when standing at different points of continuous circular stone walls. The lack of a continuous wall, the spaces between stones, may have created a series of echoes rather than a whispering gallery caused by the coherent transmission of sound around a curving stone wall as found at St Paul's Cathedral in London or the Baptistry at Pisa. These and other circular stone sites of ritual significance were investigated for comparison, and several were found to have well-known acoustic effects. A circular stone building will be likely in most cases to have some singular acoustic effects.

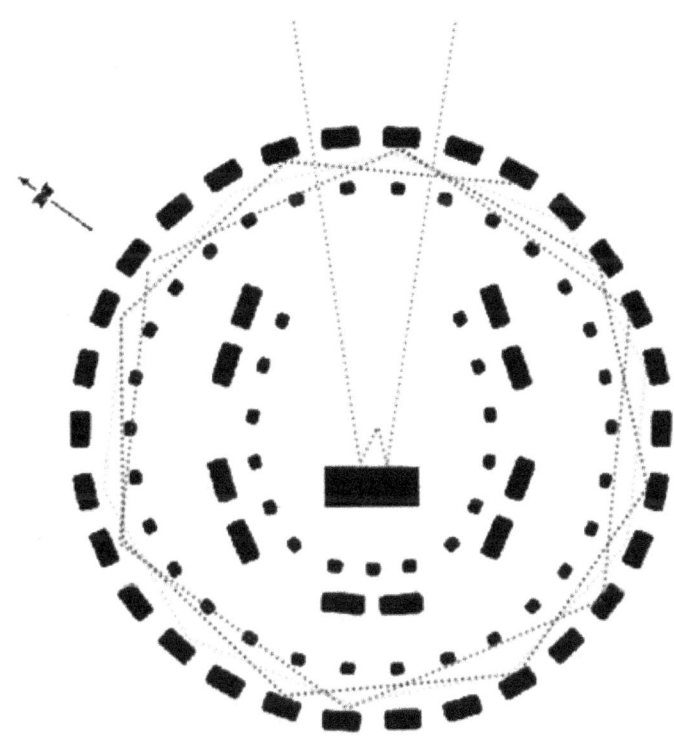

FIGURE 4.1: THEORETICAL CONSIDERATION OF SOUND REFLECTIONS IN STONEHENGE

Fig. 4.1 shows how a sound made at the edge of the circle would have reflected off each fifth, fourth and third stone as it moved around the circle. Each reflection would change the quality of the sound, and also delay its travel, causing sound to arrive later than sound travelling directly. The sound travelling from the bottom of Fig. 4.1 to the top of the picture, around the walls of the circle, would arrive later than that travelling straight upwards. This is likely to result in reverberation, or perhaps a series of discrete delays or echoes sounding somewhat like a galloping horse. It could otherwise produce a chorusing effect, depending on how long the time delays were. Fig. 4.1 also shows how the Altar Stone in the centre of the space may have acted to project the sound of someone in the centre towards the entrance of the space and down towards the avenue, the ceremonial approach to the circle beyond its bounds.

Similar calculations illustrated that the centre would act as a particular acoustic focus, reflecting sound back from all directions. Moving to the exact centre of the space would result in the sudden activation of the acoustic, giving a radically different result from that experienced only perhaps a metre away. From this central position, the trilithons seemed to be shaped as much like a megaphone as a horseshoe, focusing and projecting sound towards the avenue. (The trilithons each consist of three stones, a large pair of upright stones with a stone lintel joining them at the top. They are the tallest stones in the space.) The bluestone horseshoe seemed to have a similar effect. There was a clear acoustic orientation, a 'front' and a 'back'. The entrance from the avenue was a clear acoustic focus (front), where there were more sonic effects. The Heel Stone, Slaughter Stone and Portal Stones, some of which were in line with the avenue and entrance to the stone circle, would be likely to produce interesting echo effects, and to reinforce acoustic effects such as resonance.

It was clear from the shape of the space and materials used that there would be likely to be acoustic resonance, echo and reverberation. Because of the large size of the space, the resonances and standing waves might produce a low humming and/or echoes. One could perhaps have stimulated these resonant frequencies of the space either by playing rhythmic music, or by working in a manner that made a noise, such as working stone rhythmically to shape it, especially if one played in time to the echoes in the space. I theorised that there would be strong echoes in the space, using basic calculations based on the speed of sound and distances such as the diameter of the stone circle and distance to the Slaughter Stone. For example, sound would leave a percussive noise made on one side of the stone circle, bounce off the opposite wall, and return later as an echo. Knowing the speed of sound, and the distance taken, we can calculate the time taken, the delay time, and its associated frequency.

Speed of sound $V = 344m/s$ = distance/time
Distance from one side of the stone circle to the other and back = 33m x 2
Time delay = 66/344 = 0.19s
Frequency of echo at edge of sarsen circle = 5.2Hz
5.2Hz is equivalent to a quaver rhythm within a crotchet tempo of 156bpm
Frequency at centre = 10.4 Hz, equivalent to semiquavers within a crotchet tempo of 156bpm, or 0.1s

I encountered no mentions of such echoes in the existing literature. In addition, it seems that at the edge of the stone circle there may have been echoes at half the tempo (easier to hear and play along to) of those in the centre. At the central position, the echo may therefore have doubled in speed.

Such echoes can also appear as resonance in the space, either at this very low frequency or at octave multiples or frequency doublings. The kind of resonance that could be developed in the space might be the equivalent of playing a very slow roll on a kettledrum or some other large drum to make it ring. Another analogy is the flute-like sound that can be made by blowing over the top of a bottle. In each case the movement of air is constrained and controlled, by the drum skin and body, or the glass of the bottle, and air is stimulated into movement by hitting the skin or blowing across the bottle top. At Stonehenge, the sarsen stone and bluestone circles and their lintel rings would constrain the air, which could be stimulated by human sound, or even perhaps by the wind. As Hardy had suggested, there might indeed be a low hum or booming sound, caused by the primary circular or cylindrical resonance of the space.

These kinds of frequency, below 20Hz, are sometimes called 'infrasound' and often described as being below the range of human hearing. However, this is a little misleading. As Leventhall points out, 'sound remains audible at frequencies well below 16 Hz. The limit of 16 Hz, or more commonly considered as 20 Hz, arises from the lower frequency limit for which the standardized equal loudness hearing contours have been measured, not from the lower limit of hearing' (130–37). It would be heard by human ears if it were at a high enough volume. Sound is not inaudible below 20Hz, but has to be increasingly loud for us to hear it, at 10.4Hz 97dB and at 5.2Hz as much as 106dB. Older people can hear these low frequencies more easily, just as the young can hear higher frequencies. Loud percussion noises could reach these dB levels. Such low frequencies are often heard in the form of a rhythmic echo, rather than a sustained pitch or note. If one clapped one's hands, a 10.4Hz echo would provide an effect a little like a simple tapping rhythm at the speed of a semiquaver pulse within a crotchet tempo of about 156bpm.

The spaces between the lintels and uprights of the sarsen stone circle and trilithons, and perhaps the bluestone circle if it had lintels, could be significant resonant spaces and produce unusual resonant effects. The outer bank would be significant in that it would contain the sound within the space. The floor material would be important: whether it was chalk, mud or grass could make a significant difference. Acoustic effects would be likely to be most pronounced at the centre. There would also be echoes as sounds made returned to the centre of the space, bounced off the stones outside the circle (Heel Stone, Slaughter Stone, Station Stones), or reflected around the circle. It appeared that the main echo would have a delay time of around 0.2s, a frequency of around 5.2Hz. This might be heard at 10.4Hz (0.1s), depending on where the sound reflected, and related low frequencies might be heard.

Watson's results (2006) had indicated that sound would be contained within the stone circle, creating a sense of envelopment within the space and a change in sound when entering. Theoretical analysis had indicated the presence of resonance, a low hum possibly, echoes, reverberation, and particular effects at the centre. The next stage of the process was to use a model to in order to explore further.

The Maryhill Monument replica of Stonehenge
Models are often used within acoustics in order to conduct acoustic testing before a building is constructed, or when work needs to be carried out in a laboratory. Scale models can be problematic as they scale down acoustic effects, and other factors also have to be scaled, such as absorption by materials. We became aware of a full-size model of Stonehenge in Maryhill, Washington State, USA (Fig. 4.2, colour). It was based on archaeological plans and completed in 1926 as a war memorial. Although it was made of concrete, it was the most accurate model available, and it was decided to travel to Maryhill to carry out field tests to try to find further evidence for the acoustic effects we found through theoretical analysis.

The concrete used in the monument's construction had been polluted by salt water in a flood and was quite porous. The shapes were all more regular than the original. It had the large central Altar Stone lying down rather than standing up. It also had surfaces that were deliberately roughened to make the site look old. All of this would minimise acoustic effects. We were happy that this was a better situation than acoustic effects being exaggerated, as any results would be conservative. It was certainly far more accurate than any model we would be able to build ourselves. Concrete is still a hard reflective surface and would act in a manner similar to stone, in particular in terms of patterns of reflection, echoes, reverberation and acoustic response. This was later supported by digital modelling results. The level and intensity of effects would be reduced, but their nature should be the same. The replica stone circle itself was reasonably accurate, but the space outside was very different. The Slaughter Stone was closer than it should have been to the monument because it is sited next to the Columbia River Gorge which drops away dramatically to the side of the monument. We therefore focused on measurements inside the circle. No archaeologists had ever used the site to test theories about Stonehenge in any way, and the Maryhill Museum, which manages the site, was generous in allowing us to carry out acoustic field tests.

For these tests we used WinMLS acoustic testing software, a dodecahedron omnidirectional loudspeaker, a large sub-bass speaker, a B&K small diaphragm condenser microphone for critical measurements, a Soundfield ambisonic 3D microphone for immersive and subjective recordings, a decibel meter, and a laptop with digital audio workstation and professional soundcard for recording. (Whereas most loudspeakers point in a particular direction, omnidirectional loudspeakers project sound in all directions at equal levels; they usually have limited low frequency characteristics, but the sub-bass speaker allowed us to explore low frequencies accurately.) We took measurements using a calibrated loudspeaker system playing slow sine sweeps in various source positions. The main source positions were in the centre of the space, in front of the largest trilithon, and on the right hand side of the space (when facing the largest trilithon). We used a large number of receiver microphone positions for the measurements. We took measurements every metre in a straight line measured on an axis from the centre of the largest trilithon through the middle of the space out of the entrance and towards the Heel and Slaughter Stones. This repeated some of Watson's measurements, for we were hoping to confirm some of his findings. We also made measurements across the circle, from side to side, at a 90-degree angle to the other axis of measurements.

We also played a number of audio source files into the space and subjectively evaluated how they sounded. We had the opportunity to listen to and record a flautist, dijeridoo player and actor in various positions and while moving around, and to evaluate the effectiveness of speech and musical sound. We experimented with using sine and square waves at single frequencies, slowly swept upwards manually, to investigate resonances within the space. We also used rhythmic pulses and waveforms (square waves, because they were easier to hear at low frequencies than sine waves) at various tempi and frequencies, to explore frequencies of 1–20Hz. We measured variations in loudness caused by resonance by making recordings and using a handheld decibel meter. We explored echoes by using impulses, making short sharp sounds so that we could hear, record and evaluate echoes.

Our measurements found a reverberation of about 1.5s (T30 = 1.5s). EDT was about 0.8s. There was quite a large amount of background noise, around 30dB, and so figures below this level were unreliable. There was a considerable amount of wind noise, of which unfortunately we took little notice, especially considering the information on Hardy and wind noise discussed already in this report but discovered in reality after the Maryhill trip. The wind noise was very loud in the space, however, especially as the day continued, and eventually it made measurements of any sort impossible. This was largely due to the positioning of the monument next to a river gorge, but in retrospect the wind seemed louder inside the monument than outside, and it is possible that Hardy was right, and that the sound of the wind whistling through the stones was (and is) an important feature of Stonehenge.

The reverberation time is shown in Fig. 4.3. The Schroeder curve shows stepping (undulations up and down), an indication of the presence of resonances and/or echoes. We were able to discover and map variations in loudness, and to create powerful standing wave resonances using both single low frequencies and musical percussion rhythms between 0.5 and 15Hz. This replicated the results of Watson's pilot study (2006).

The impulse response of the space shows a number of reflections and echoes, although these were particularly localised (Fig. 4.4). We found in fact that one could hear a clear echo when one stood in front of the largest trilithon and could see the Maryhill equivalent of the Slaughter or Heel Stone. This lies in a straight line between what would be the ceremonial approach to the site, the 'entrance', and the centre of the circle. Of course

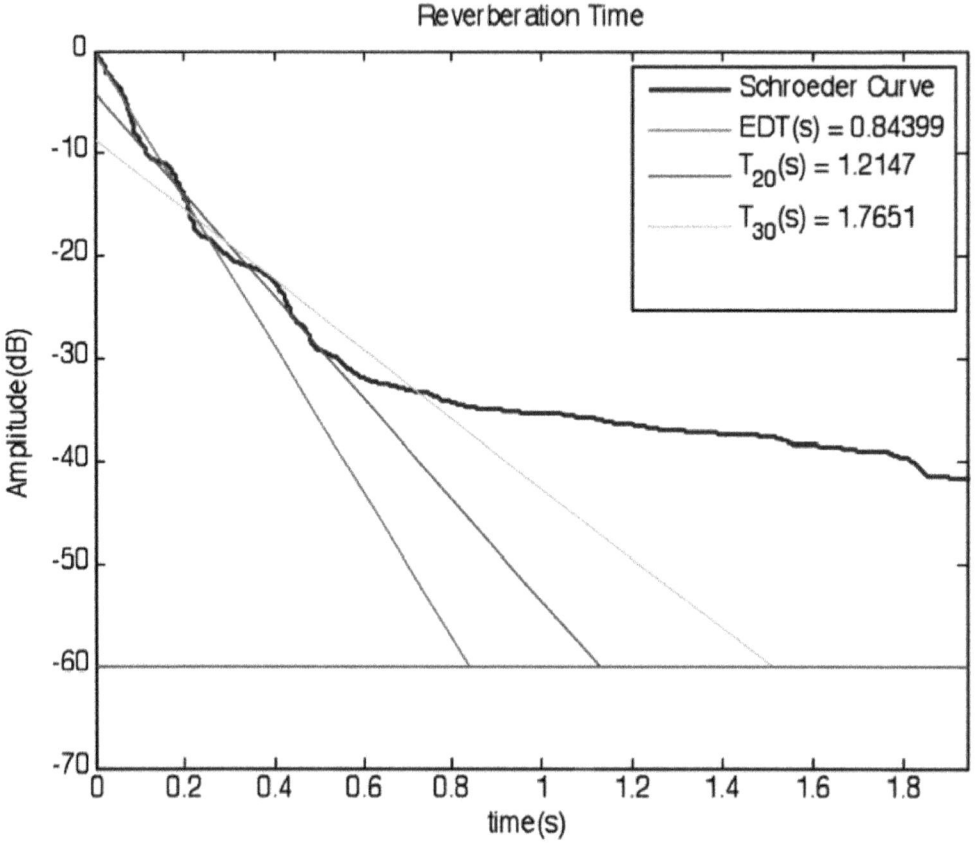

FIGURE 4.3: REVERBERATION TIME AT MARYHILL

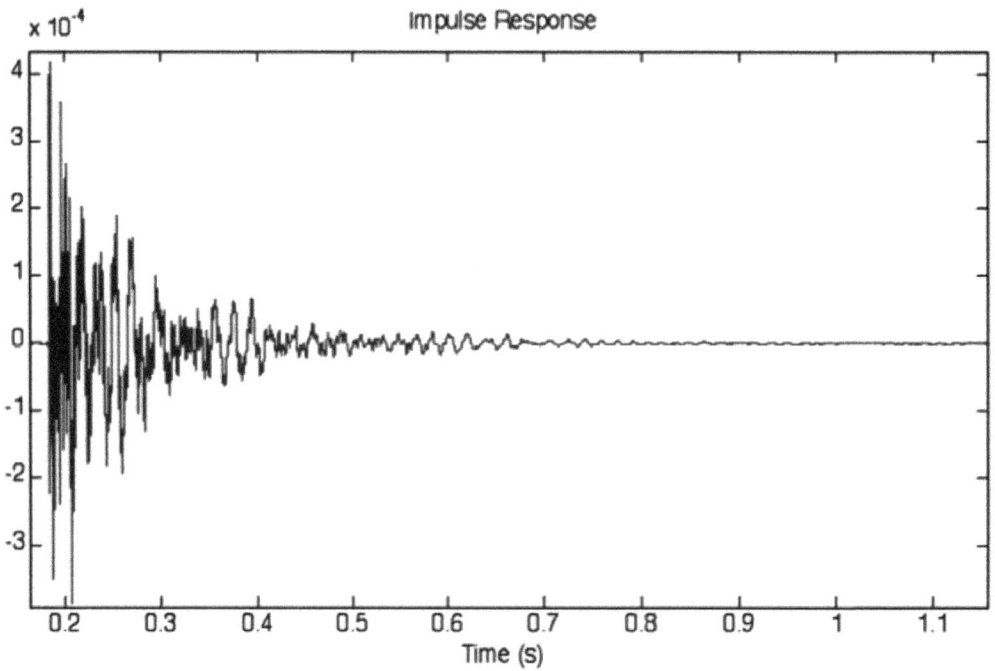

FIGURE 4.4: IMPULSE RESPONSE OF MARYHILL

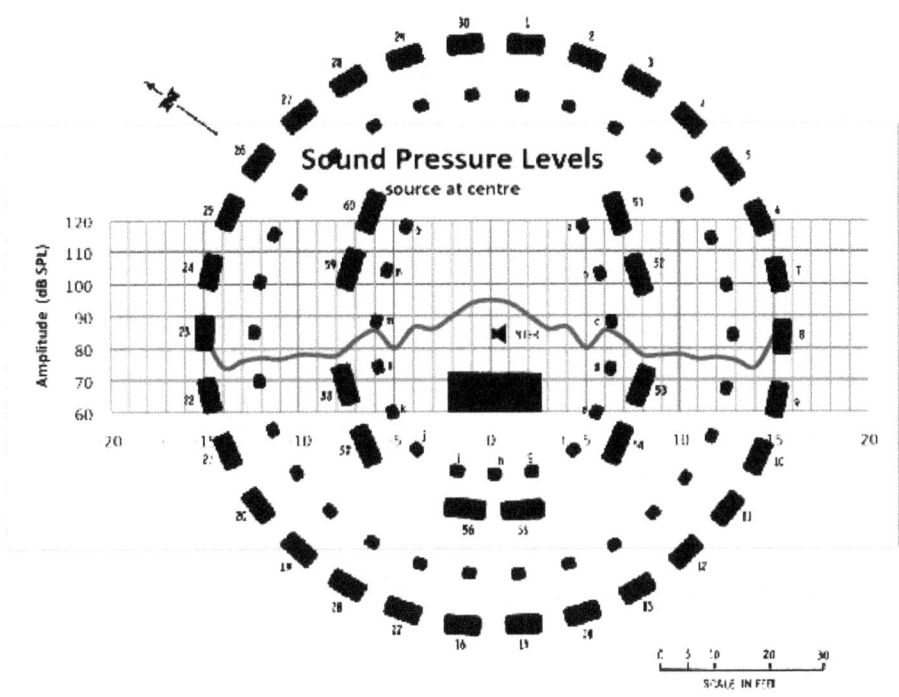

FIGURE 4.5: RESONANCE AT MARYHILL

there are no Station Stones or other stones outside the circle at Maryhill. I found later that at Stonehenge there are similar echoes that come from the Station Stones, and that the main echo in the site now comes from the Heel Stone. It seemed that the external stones worked as echo stones, catching sound and projecting it back into the circle. The placement of these was vital, and carefully aligned with acoustic and visual sightlines.

Subjectively, there was a powerful sense of envelopment and inclusion within the concrete circle. One felt 'inside', separated from the outside world. Sounds from outside seemed distant and disconnected. Within the circle, musicians and the actor immediately stood on the Altar Stone, in the centre. Perhaps rather than standing, this was a prone 'stage stone'. They noticed that the centre was the most flattering position to play music. Recordings of music were also pleasantly flattered within the space, rather than being disrupted or blurred by its acoustic. We played recordings of music from various world traditions that featured voice, percussion and clapping, in order to simulate possible group musical activities in the space. A pleasant reverberation was added to them, without any severe muddying or colouring of the sound.

Speech was diffuse and intelligible in the space, and clarity was good. One could talk to another person on the other side of the space quite easily, even if they were behind a stone. When walking behind a stone, speech bounced off other surfaces and was still audible: the stones did not seem to reduce the level of sound. This allowed one to stand behind a stone and make it appear the stone was speaking. Being exactly in the centre of the space amplified perceived sound, and this was definitely a position with special acoustic properties. When entering the circle, standing underneath the lintels of the circle had a very pronounced effect. It felt as one entered the circle as though one were going through an acoustic door, entering another acoustic world. If one stopped under the ring of lintels, with an upright stone each side, there was a different, enclosed, resonant acoustic. However, this acoustic was different enough from either the inside or the outside to enhance the process of crossing the threshold from outside the circle to the inside. Watson speaks in some detail about this effect at Stonehenge itself. Some other positions had very different acoustic properties. The space between the bluestone circle and outer circle was quieter than other positions, whereas being right next to the outer circle was louder.

Occasionally when walking around the space one would hear a word seem to leap out of the air, or there would suddenly be a strange acoustic effect. By being in a specific position while someone else was standing in another particular place, talking in a specific direction, the acoustic would suddenly come alive. This proved frustratingly impossible to predict or recreate, but repeatedly one would suddenly hear a strange effect applied to on one's own or someone else's voice. High frequencies were affected in particular. These effects would have been stronger in the original site where the stones were more reflective, shaped to be smoother, and curved in a way that would focus sound.

However, we also found remarkable low frequency results. We were able to stimulate standing waves, resonances, so strong that sound was equally loud on each side of the circle, rather than being louder by the loudspeaker that was making the sound at the circle's edge. We made the space resonate like a wine glass or bottle, using individual frequencies generated from the computer. A short repeating drum sound, played at the correct tempo at high volume by loudspeakers, was also able to stimulate a sympathetic vibration in the space, and as one walked around, the short attack of the drum was made longer, changed into a low bass throb, more like a synthesizer than a drum, getting louder and softer as one walked across the space. Low frequency boost, patterns of loud and soft, and softening of attack are all typical effects when modal acoustic resonance is present.

Fig. 4.5 shows the volume level increasing as one gets to the edge of the circle. A graph of volume is superimposed on top of an image of the Maryhill Stonehenge. One can see, for example, the correlation of the position of the bluestone horseshoe shape with an increase in volume. These and other readings were used to show some of the circular modes of vibration in the space. Sound does not resonate in a circular space like a string, in a straight line, but in circles, more like a cylinder and/or drumskin. In Fig. 4.6 (colour), actual volume levels are shown in red. These seem to be a mixture of modes 4 and 10 (shown by yellow and green lines). There are similarities between the red and green lines, except in the centre where red and yellow become more similar. By drawing blue circles lined up with the peaks of the green line, one can see a clear correlation with the positions of the stones. These circles point out how the positions of the circles and horseshoes of stones at Maryhill (and originally at Stonehenge) act to force particular resonant behaviour. Analysis of results is ongoing, and detailed results from Maryhill will be part of a future paper focusing further and specifically on the work at Maryhill. There were many other interesting findings. One was the movement of shadows within the space as the sun rose. (We were there shortly after the solstice. The space is aligned astronomically in as similar a way as possible to Stonehenge, considering the difference of geography. It is aligned to the passage of the sun at the solstices, as well as to various sunrises and sunsets and the rising and setting of the moon.) Another was the nature of the relationships to the monument of local people and visitors that had developed. Indeed, the Maryhill Monument is interesting to study in itself. For the past 83 years it has acted much like a running project in experimental archaeology using a living model of Stonehenge. If nothing else, the research visit to Maryhill provided an interesting phenomenological, experiential, partial understanding of what it might have been like to travel to and experience Stonehenge.

To summarise results at Maryhill, there were a number of acoustic effects present. We measured a reverberation time T30 of about 1.5s and early decay time of about 0.8s. It was possible to stimulate the modes of resonance in the space using single frequencies and rhythms. These caused highly localised differences in volume and unusual sonic effects. Low frequencies in the space were very interesting and require further study. Wind noise was very loud in the space. The results of Watson's pilot study were supported, in terms of changes in sound when approaching and entering the space and variation in volume within the space. The axis of the monument that corresponded to the ceremonial approach to the space (and the break in the bank) at Stonehenge was of particular interest, and sound effects seemed to be focused in that direction. We also discovered effects at 90 degrees to this axis, going across the space between the two pairs of equal-size trilithons. The space underneath the lintels of the outer stone circle had a particular effect, creating an acoustic doorway and threshold and enhancing the effect of entering the interior. The acoustic outside the stone circle was different from that inside, and there was a subjective sense of aural envelopment, enclosure, and separation from outside when within the space, one that grew as one spent more time there.

The central area, contained by the horseshoes of trilithons and bluestones, from the Altar Stone to the 'entrance', had a different acoustic from the rest of the space, and seemed most sonically active, having the clearest acoustic effects. There was an increase in volume when close to and inside stones of the outer circle, and when close to and in front of the trilithons. There were a number of acoustic dead spots where sound was quieter, in particular halfway between the outer sarsen circle and the bluestone circle inside, and also behind the middle-size trilithons. There was a clear front and back to the space. The Altar Stone if laid flat would have made a good stage for projecting sound as well as aiding sight lines. Speech intelligibility was subjectively assessed as good across the space, and standing behind stones did not seem to mask the voice. The reverberation provided a pleasant, flattering acoustical quality for both speech and various kinds of musical material. There were echoes in the space, only heard when in front of the largest trilithon, and looking towards the Heel Stone. The echo was at its strongest at each end of a line connecting these two points. We were later able to return to Maryhill and further explore the acoustics of the space. Results will be published in the near future.

Computer acoustic analysis

A digital model of Stonehenge 3vi was obtained from the Imigea digital modelling company's virtual museum. This model was then acoustically analysed using Odeon software. The model is still being analysed and will be presented in more detail in a future publication. Initial results are discussed here. The acoustic model lacked detailed figures for the reflective qualities of the stones. The model was given the reflective qualities of marble, as this was the closest material available in terms of acoustic properties. The model was tested with rough concrete surfaces, and similar results were produced,

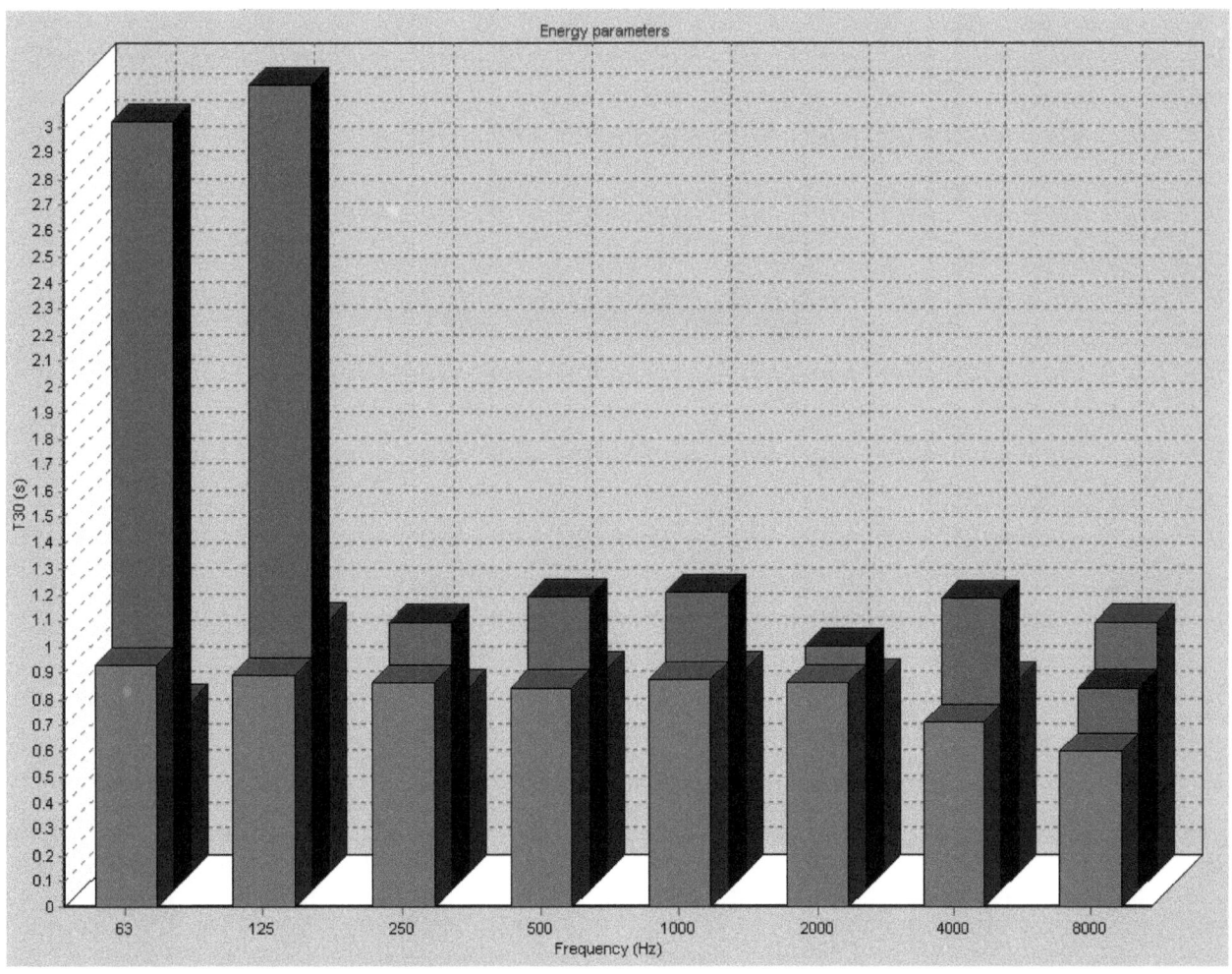

FIGURE 4.7: REVERBERATION TIME T30 AT DIFFERENT FREQUENCIES

		63Hz	125	250	500	1000	2000	4000	8000
EDT	(s)	3.93	3.88	2.27	2.56	2.36	1.87	1.8	0.94
T30	(s)	2.47	2.51	1.05	1.10	1.01	0.68	0.68	0.70
SPL	(dB)	15.4	15.5	8.7	9.4	8.8	4.0	2.4	-1.8
C80	(dB)	-6.8	-6.7	-3.0	-5.7	-5.2	-2.2	-0.4	5.1
D50		0.14	0.15	0.29	0.17	0.19	0.31	0.41	0.70
Ts	(ms)	234	233	152	185	180	135	111	55
LF80		0.132	0.138	0.158	0.255	0.277	0.292	0.299	0.315

FIGURE 4.8: ACOUSTIC CALCULATIONS FOR THE CENTRE OF STONEHENGE

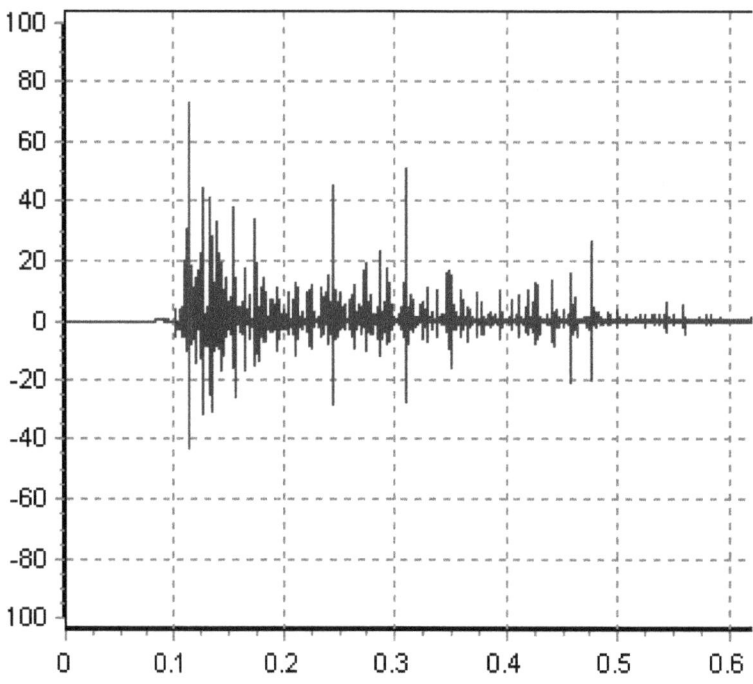

FIGURE 4.9: BINAURAL IMPULSE RESPONSE

which indicates that the choice of which hard reflective surface to use was not critical for these measurements. It also indicates that the model at Maryhill has some validity despite its concrete construction. Fig. 4.7 shows reverberation details for the model that vary between different frequencies and in different positions.

The acoustic model uses ray tracing, and is therefore likely to be inaccurate at very low frequencies. But with a substantially larger reverberation than other positions and frequencies shown at 125Hz at the edge of the circle, at this frequency the order of difference is significant, and results will be at least partially accurate. Here the reverberation time is 3.2s, compared to 0.9s inside the circle at the same frequency. This shows a strong low-frequency resonance probably caused principally by the outer sarsen ring of uprights and lintels. It also shows that this low ringing would happen far more strongly at the edge than anywhere else.

Fig. 4.8 shows the powerful acoustic effects at the centre of the space. Sound leaving a source reflects back to the centre from all directions, bouncing off the stone circle and returning. One can see that reverberation and early decay time are much higher at low frequencies (as at the circle edge). Volume levels (SPL) are also higher at the edge. Clarity (C80) and definition (D50) are correspondingly higher at high frequencies, the reverberation making clarity worse at low frequencies.

The impulse response in Fig. 4.9 exhibits a number of spikes after the initial sound, showing echoes in the space. A reverberation with no echoes would have a more smooth curve. Mapping of clarity (C80) in the space has shown that there are lines of clear sound, much like lines of sight, along which sound transmits. Particularly interesting was the way the sound was transmitted out of the entrance down the ceremonial approach to the site. Sound escapes the outer bank at Stonehenge in the Odeon model only at specific points; these are illustrated by the dark lines in the speech transmission index map shown in Fig. 4.10 (colour; the map for clarity is similar). The line leaving at the top seems to head off towards the Cursus barrows. On the right there are four lines. The lowest may aim towards Durrington Walls or the Cuckoo Stone, the next upwards is aimed straight at the space (or stone) next to the Heel Stone. The next aims towards the western end of the Cursus. It is interesting to consider whether sound from Stonehenge could have been heard in these places; this warrants further investigation. Extremely low frequencies can travel particularly long distances.

LG80 is a measure of how much an acoustic gives a sense of envelopment in any specific position. The map of envelopment (Fig. 4.11, colour) shows clearly how, within the stone circle, LG80 is higher (red) than that outside, as high as 21.6 near the centre. This seems to illustrate the social stratification of the space. In concert hall environments LG80 is used to identify the seating positions where tickets are most expensive. The positions with the highest figures for envelopment are the most desirable and significant, as Soulodre *et al* explain:

> Spatial Impression is known to be an important part of good rated concert hall acoustics and it is now well established that spatial impression is composed of at least two components: apparent source width (ASW) and listener envelopment (LEV). ASW is defined as broadening of the apparent source width of the sound source, while LEV refers to the listener's sense of being surrounded or enveloped by sound.

(ASW equates to LF80, as seen in Fig. 4.8. LEV is described by LG80.)

FIGURE 4.2: THE MARYHILL MONUMENT, MARYHILL MUSEUM OF ART, WASHINGTON STATE, USA (AUTHOR'S PHOTO)

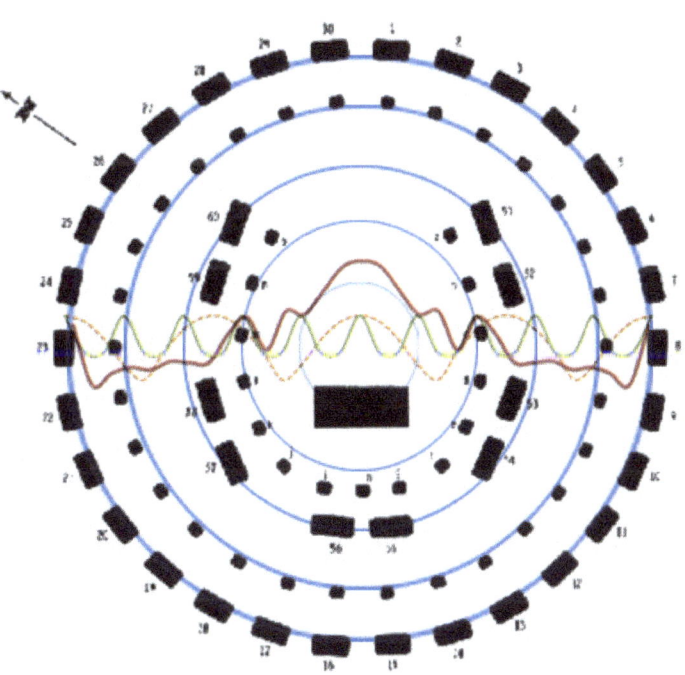

FIGURE 4.6: REAL AND THEORETICAL MODES AT MARYHILL

The Sounds of Stonehenge

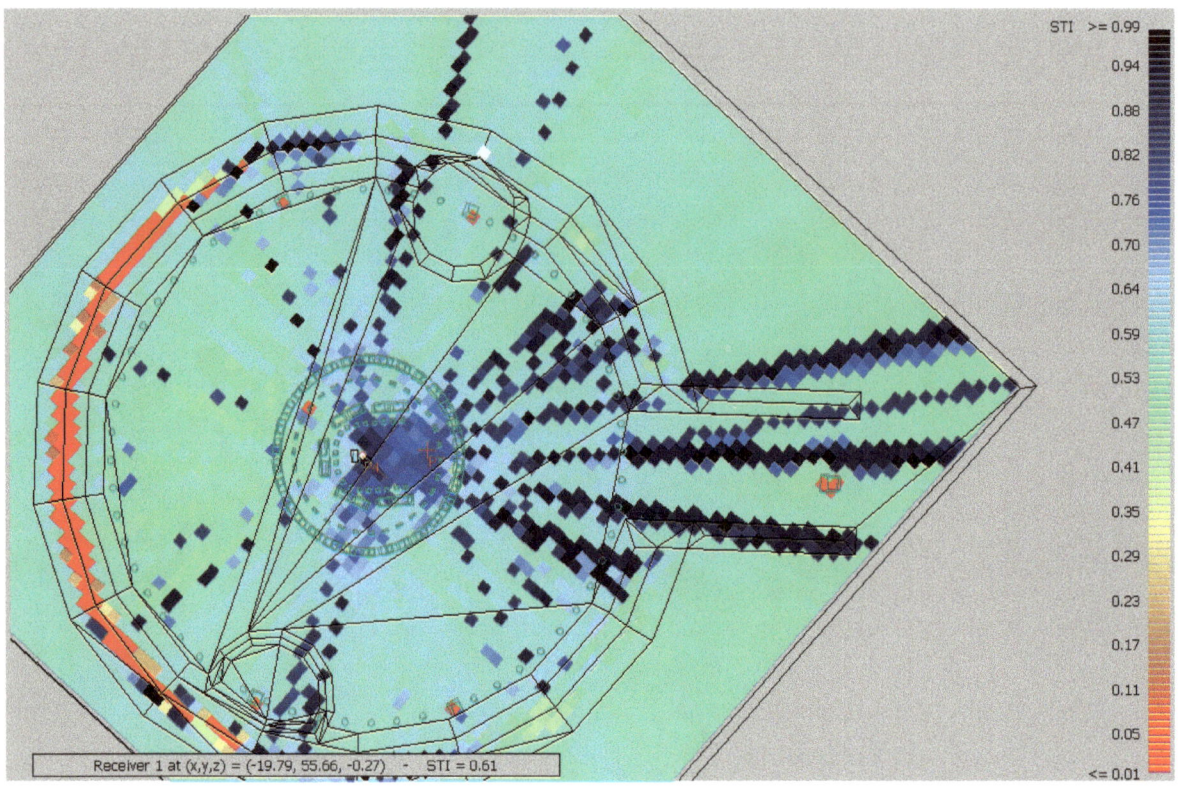

FIGURE 4.10: SPEECH TRANSMISSION INDEX ABOVE 0.9 AS INDICATED BY BLACK LINES

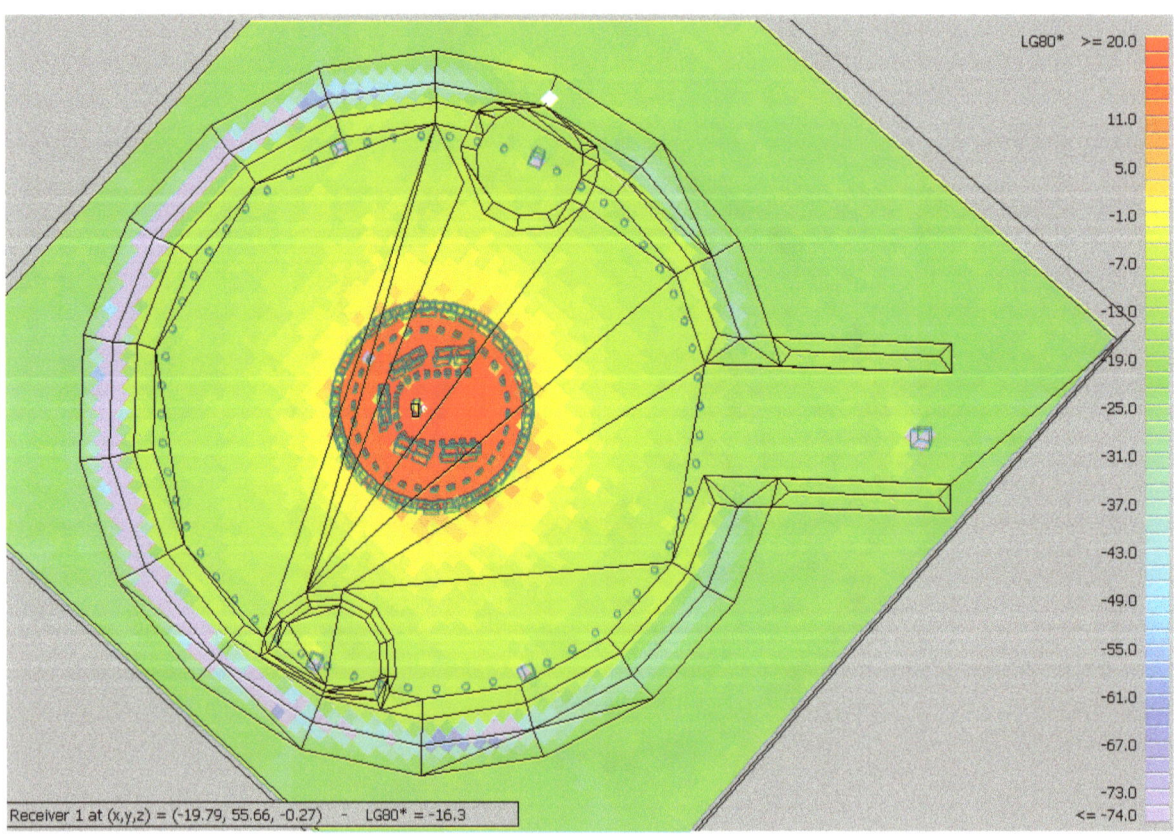

FIGURE 4.11: LG80 ENVELOPMENT MAPPED OUT IN THE SPACE

Subjective listener aspect	Parameter	Range (AkuTEK)	Music	Stonehenge	Oslo	Elmia	Vienna
Level of sound	G/SPL	-2 to +10dB	>3dB	8.8dB	1dB	4dB	3dB
Reverberance	EDT	1 to 3s	2.2s	2.36s	1.5s	1.7s	1.9s
Clarity	Clarity C80	-5 to +5dB	-1 to 3dB	-5.2dB	2dB	1dB	2dB
	Definition D50	0.3 to 0.7		0.19			
	Centre time Ts	60 to 260ms		180ms			
Source width	LF80	0.05 to 0.35	>0.25	0.277	0.23	0.18	0.20
Envelopment	LG80	-14 to +1 dB		6dB	-8dB	-5dB	-3dB

FIGURE 4.12: COMPARATIVE ACOUSTIC VALUES

	P1	P2	P3	P4	P5	P6	P7	Heel	O/E/V	R
SPL(A)	36.5	42.1	42.2	29.9	7.6	21.1	17	22.1	**1 to 4**	-2 to 10
T30	**0.98**	**0.86**	**0.85**	**0.91**	**0.87**	**0.81**	**0.98**	1.01		0.8 to 1
EDT	0.64	0.54	0.50	**1.34**	**2.0**	0.04	**1.12**	**1.24**	**1.5 to 1.9**	1 to 3
STi	**0.74**	**0.74**	**0.81**	**0.54**	0.49	0.99	**0.54**	0.58		
LG80	16.4	18.8	17.5	16.5	-12.8	-15.6	**0.8**	10.4	**-8 to -3**	-14 to 1
LF80	**0.074**	**0.099**	0.035	**0.284**	0.029	0.001	**0.196**	0.491	**0.18 to 0.23**	.05 to .35
C80	10.2	13.1	13.7	**-0.4**	**0.4**	19.0	**4.2**	**0.8**	**1 to 2**	-5 to 5
D50	0.87	0.92	0.92	**0.33**	**0.4**	0.98	**0.56**	0.37		0.3 to 0.7

FIGURE 4.13: ACOUSTIC VALUES IN DIFFERENT MEASUREMENT (LISTENING) POSITIONS (CENTRAL SOURCE AT 1kHZ, VALUES WITHIN SUGGESTED RANGE R IN BOLD)

Typical ranges of acoustical parameters for concert halls (see AkuTEK) are indicated in Fig. 4.12, along with appropriate values for musical source material where available. Skålevik provides comparisons with results produced by Odeon for digital models of existing concert halls (Oslo, Elmia and Vienna). Odeon also provides guidance in its manual for recommended figures for symphonic music (see the column 'Music' in Fig. 4.12). It is interesting to compare these with figures provided by Odeon for the Stonehenge digital model. Stonehenge is loud; it has figures for reverberance and source width for musical sounds that are as good as the three concert halls. It has poor clarity and definition but acceptable speech intelligiblility, and envelops the listener in an extreme fashion.

High transparency or speech intelligibility is indicated by low values of the centre time Ts (Kuttruff: 211). According to Rumsey, subjective 'clearness' relates closely to the measured concert hall acoustics parameter D50 (definition or *Deutlichkeit*), and is an indication of direct to reverberant ratio. D50 (intended for speech) above 0.5 is 'excellent', and C80 should be above 0, aiming for 15, which is good. Although the averaged figures for D50 and C80 at the centre of Stonehenge are not good compared with those for the concert halls, values vary at different positions and are good in places.

In Fig. 4.13, the results are as at positions 1, 2, 3 etc (P1, P2, P3 etc), plotted in Fig. 4.14. There is also a set of results with the source at the Heel Stone instead of the centre, where it is for P1–P7. O/E/V is the range of values in the Oslo, Elmia and Vienna concert halls. R is the AkuTEK suggested range of values from Fig. 4.13. Values of STi (speech transmission index) of 0–0.3 are bad (subjectively), 0.3–0.45 are poor, 0.45–0.6 are fair, 0.6–0.75 are good, and 0.75–1 are excellent (Christensen: 7–86). T30 (for rock music) should ideally be 0.8 to 1s. EDT for speech should be 0.3–1.2s, for orchestral music 2.2s (Larsen *et al*: 4; S von Fischer: 3).

Fig. 4.13's results show that the space is louder than a concert hall, and very loud for an outdoor space. T30 across the space, and even when standing at the Heel Stone, was close to ideal (at least close, appropriately enough, to the concert acoustic requirements of amplified rock music). Most of the results (P1–P7) are based on software modelling with a sound source placed at the centre of the space. When a source is placed at the Heel Stone and readings are taken in the centre, results

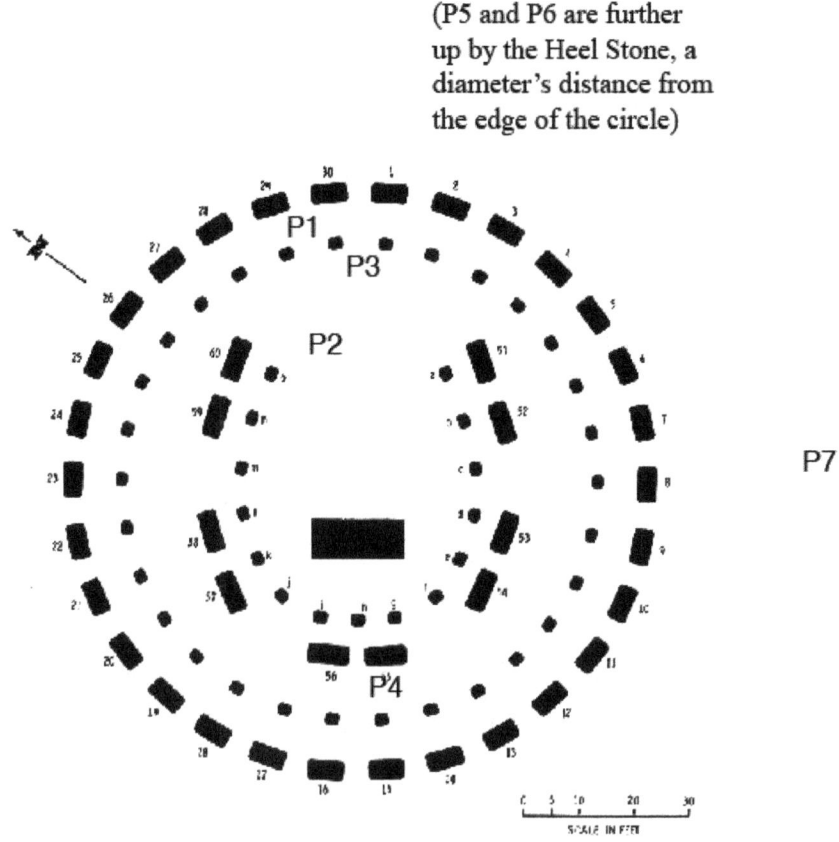

FIGURE 4.14: RESULTS POSITIONS P1–P7 FOR FIGURE 4.13

are mostly good, better for speech than for music. Thus if sound were made at the Heel Stone with the intention of it being heard in the centre of the circle, the acoustic properties of the space would support speech better than music. Figures of T30, reverberation time, are fairly uniform, in all positions between 0.8 and 1. This is within the ideal range for rock music. This is significant, since it shows that the reverberation at Stonehenge would naturally fit loud, rhythmic music. Quiet, or orchestral, or harmonically rich music is generally performed in a venue with a longer reverberation time, reflected in larger values of both T30 and EDT, perhaps 1.5–2.2s. We are also used to hearing vocal music in highly reverberant spaces, the acoustics of churches being designed to provide long reverberation for singing. Hearing the reverberation tail of the previous note allows a singer to pitch the next note in tune, and is generally considered to flatter the sound of the voice. This is because it helps to hide the small intonation (pitch) inaccuracies and variations that are a natural part of vocal production by mixing them with the direct sound, producing a chorus effect which reduces the accuracy of the listener's perception of the pitch of the sung note.

This is true to such an extent that large reverberant spaces are routinely added to dry popular music recordings made in studios using electronic reverberation units.

It is not helpful, however, for loud, rhythmic music to be performed in highly reverberant spaces. This is because percussionists need to hear the exact timing of the previous note(s) in order to know the exact moment at which to play the next rhythm. Thus concert spaces for popular music tend to have relatively dry acoustics, with EDT or T30 about 0.8–1. We can see from the Odeon results that the acoustic in the centre of the circle would make speech clear, well-defined and intelligible, especially in the central area enclosed by the trilithons and the entrance to the space. We can also see that if there were music in the space, singing voices and pitched instruments would not be as well supported as rhythmic music. Singing or instruments would have sounded better to those outside the circle, especially at the Heel Stone, where there would have been particularly strong musical enhancement and acoustic effects. Sound made at the centre would have been very much enhanced acoustically at the Heel Stone, and sound made at the Heel Stone would have been enhanced, though not quite as much, when heard in the centre, speech being better transmitted in this case.

STi at the side of the Heel Stone (P6) was excellent, as was clarity and definition. Impressively, these figures are substantially higher than if standing in the central part of the circle. There is also a considerable difference compared with standing 1m to the right of the Heel Stone. The Heel Stone was originally one of a pair (J Richards: 18), and it seems that sound was aimed in particular either at this missing stone or the space between the two. Further research will investigate the effect on the acoustics of including this other stone in the simulation.

Also remarkable are the figures for D50, STi and C80 in the centre of the space. They are considerably higher than the concert hall figures, and it is clear that speech would have been very clear in the central area of the space. Had the space been small, unreverberant or lacking in obvious acoustic effects, like a modern living room, this would be less unusual, but such is not the case. Within the stone circle there are very high figures for envelopment, substantially higher than the figures for the concert halls, and it is clear that those within the space would feel enclosed and enveloped by it.

In addition, Odeon allows one to produce auralisations, sound examples, such as someone clapping or singing, for hearing how audio would have sounded in the space with the listener positioned at a given point. These auralisations illustrate a distinct reverberation that changes in tone and character in different positions. Odeon also shows how the sound travels around the space and arrives at the receiver (listening) position. Fig.4.15 illustrates this, and shows how the sound arrives directly and indirectly at the listener, producing reverberation and listener envelopment. Results varied in some cases metre by metre, with the acoustics depending on precise position of listener or sound source, much like Maryhill. Indeed, results agreed with and largely confirmed those from Hardy, Watson, theoretical investigation and Maryhill.

Conclusions

Information about music and culture informed what sounds we should expect to find at Stonehenge. We could expect music that is part of a ritual activity and addresses the supernatural; is integrated into its social context; is combined with dance; is designed to lead to the achievement of altered states of consciousness or trance; or that changes the ambience of a gathering. It may be used both to bond the community and to establish one's position within the community. It may involve communication with gods and the ancestors, may be part of rites of passage from birth to death, and may be used as part of healing of some sort.

Hardy had noticed a sense of enclosure and envelopment in the space, a reverberant acoustic, echoes when entering the sarsen circle, interesting sound at the centre, and a low booming hum that was caused by the wind. Watson's pilot study had shown that sound would be contained within the stone circle, creating a sense of envelopment within the space and a change in sound when approaching and entering from the avenue. Theoretical analysis had indicated the presence of resonance (agreeing with Watson), as well as a low hum, echoes, reverberation, and particular effects at the centre (agreeing with Hardy). It indicated that echoes and resonance at the centre of the circle may be at be heard as semiquavers within a 156bpm crotchet tempo (equivalent to 10.4Hz) or as quavers (at 5.2Hz) at the edge.

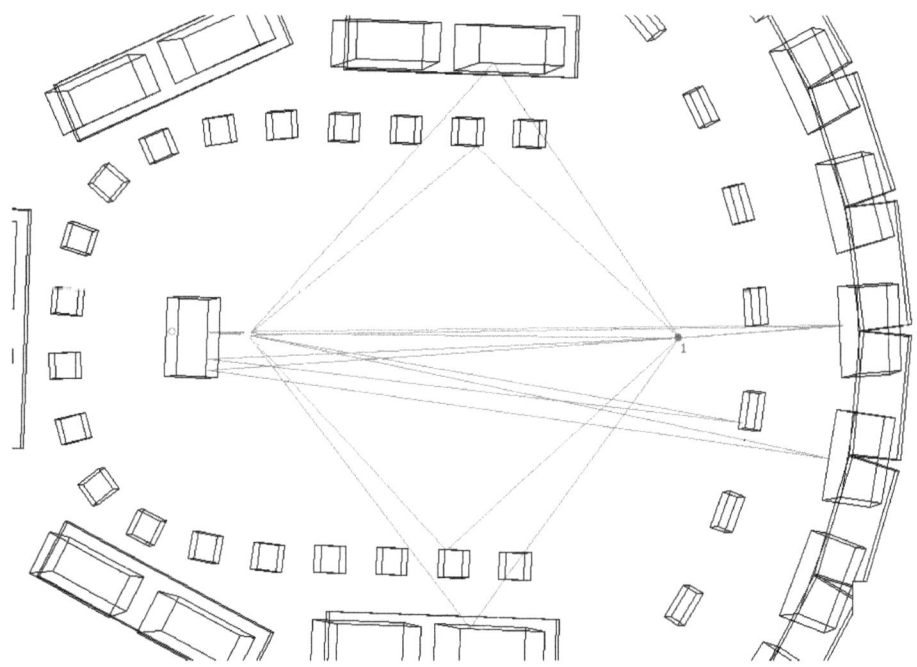

FIGURE 4.15: REFLECTIONS FROM A CENTRE SOURCE TO RECEIVER (RIGHT)

Acoustic analysis of the Maryhill Monument concrete replica of Stonehenge, utilising it as a full-size model, gave indications of the acoustic behaviour of Stonehenge. It indicated reverberation time of 1.5s (T30) and 0.8s (EDT). The site was able to resonate at a number of frequencies as a response either to generation of specific frequencies or to simple rhythmic musical sequences at tempi equivalent to infrasonic modal low frequencies of vibration. A focus of acoustic effects was observed along what appeared to be the main acoustic axis of the space, leading from the Heel Stone through the 'entrance' from the avenue in the northeast, to the centre, Altar Stone and largest trilithon. There were also acoustic effects along an axis across the circle at 90 degress to this central axis.

The outer stone circle created a sonic threshold, with an unusual acoustic effect under the circle of stone lintels, and a marked difference in acoustic inside and outside the stone circle, enhancing the moment of entering. There were acoustic effects at the edge of the circle, where sound was louder. There was a sense of acoustic enclosure and envelopment within the circle. The acoustics seemed to focus on the central space bounded by the trilithons and entrance. There was considerable variation of acoustic within the space, with odd effects at specific positions, and a 'backstage' area behind the trilithons. There were echoes evident only along the main acoustic axis. Speech intelligibility, clarity and definition were subjectively assessed as good within the space, which was suitable for music or speech. Results from Hardy, Watson and theoretical work were supported.

Analysis of a digital model of Stonehenge using Odeon acoustic measurement software showed that standing at the exact centre of the stone circle, one would experience boosted low frequencies, very high listener envelopment, reverberation and unusual acoustic effects including echoes. Being at the centre would stop one from hearing the speech of other people in the circle in a clear, intelligible fashion; it seems that others would hear this person clearly, but the central person would not hear others clearly. Words spoken at the centre would be intelligible, clear and well defined to others, in particular within the central area of the stone circle and next to the Heel Stone. Sounds from the centre would be amplified, and would have acoustic effects that would change their tone without destroying speech intelligibility. Standing at a specific sweet spot in the centre would have meant that acoustic effects would be more powerful than elsewhere. This sweet spot would have been small, suited to only a single person. Singing (or musical instruments played) at the centre would be sonically flattered when heard outside the stone circle, especially next to the Heel Stone, where there would be long reverberation and echoes. This implies that there may well have been people singing (or playing instruments) in the stone circle, and that there would be people listening outside it at the Heel Stone or within the encircling bank. Rhythmic music made in the centre would sound better within the stone circle and worse (confused, lacking coherence, definition and clarity) outside.

There is an extremely high measure of perceived listener envelopment inside the stone circle. This falls away between the stone circle and the encircling bank to give comparable figures to seats increasingly far away at a concert hall. There was still a sense of envelopment and inclusion outside the stone circle while within the bank, but with clear delineation of the centre as the most high status space in terms of acoustics.

Standard measures of acoustics such as envelopment, clarity and definition, as well as speech intelligibility, are in some positions at Stonehenge better than the equivalent values for the Musikvereinsaal in Vienna in Skålevik's study, and Stonehenge had acoustic properties that seem too carefully controlled and close to ideal values in many cases to be anything other than at least partially intentional. Sound would be transmitted with very high levels of perceived clarity, out of the space along specific straight lines aiming towards the Cursus, Durrington Walls, the Slaughter Stone and/or the Heel Stone's now lost partner.

The whole of the space inside the stone circle was a place where people would have found listening an involving process. This implies that people would have been present inside the space during ritual activity. Being 'backstage' behind the trilithons would have protected listeners to some extent from acoustic effects, and made them perceive themselves as separated them from central activity. At the edge of the circle next to the sarsen stones, low frequencies were boosted powerfully and had a long reverberation time T30 of up to 3.2s at 125Hz.

Standing at the Heel Stone, there would not have been as strong a sense of envelopment, although sound made at the Heel Stone would have created a sense of envelopment when listening in the centre, much more so than sound made at other positions outside the circle. In this way the Heel Stone position would have been connected to the inside of the circle in a way that the rest of the surrounding space was not. It is unclear whether the Slaughter Stone (not modelled) would hinder or interact with this effect. Speech from the centre would have been very clear, intelligible and well defined immediately next to the Heel Stone, and much less so as little as a metre away in front of the Heel Stone itself. Musical sound made in the centre would have been enhanced by reverberation at the latter position. Small changes in the position of a person speaking at the centre could have reversed these conditions. If there were two people, one standing in front of the Heel Stone and one next to it, someone in the centre could have spoken clearly to one or the other by moving only a little to one side, using sightlines as a guide. The acoustics imply that someone at or near the Heel Stone may have spoken to those in the centre of the space, taking advantage of the particularly clear acoustic path between the two positions.

Echoes and other acoustic effects would have been heard at the Heel Stone. The acoustic significance of the Slaughter Stone, Portal Stones and Station Stones warrant further investigation, but it seems that their placement may have had particular and specific acoustic effects. Their placement may cause them to interact in particular ways with the modal resonances in the space.

Hardy's description of echoes, reverberation, low frequency resonance and acoustic focus at the centre, as well as Watson's observations of envelopment and the importance of changes in acoustic along the ceremonial approach to Stonehenge, are all backed up by theoretical analysis, field tests at Maryhill and digital acoustic modelling. There were clearly acoustic effects present that were powerful and dramatic, so strong that they could not have gone unnoticed. The results from the different elements of the project, even at this initial stage, support each other strongly.

One could never prove absolutely to a sceptic that acoustic design was intentional, but this level of acoustic interest and detail seems to make it highly likely that the acoustic properties of the space were a consideration in the development of Stonehenge. We acknowledge that the builders of Stonehenge could construct the site in a way that created visual and physical patterns that were able to map astronomical progress; why would we think it unlikely that they could also consider sonic construction? One no more needs an understanding of acoustics to create acoustic affects than one needs an understanding of the theoretical mathematics of force and energy, or architectural training, to raise a stone lintel exactly level on two uprights. When developing the site, the builders would have noticed the way the sound changed, and thus come to understand how placing stones changed the acoustics of the site. Over hundreds of years, trial and error could be a powerful method of learning. We believe the site was consciously designed in a visual sense, not the result of random placement of stones; it seems very likely the site was designed acoustically as well as visually.

However, intentionality is perhaps an irrelevant consideration. Whether the acoustics were intentional or, surely less likely, a remarkable, happy accident, the acoustic effects would have been obvious to anyone in the space, especially considering earlier comments about the importance of sound in oral cultures. It is likely that acoustic effects would have affected, and been integrated into, ritual activity in the space. Just as one might be tempted to shout at an echoing mountain or in a tunnel, the echoes in the space would have been intriguing. Just as the reverberance of a cathedral makes one quiet because speech and footsteps are amplified, transmitted around the space and sustained, being within the Stonehenge circle or at the Heel Stone would have imposed a sense of awe upon the listener, a sense of being in a special place. It is clear that the acoustics of Stonehenge and its environment are significant and worth consideration.

This has implications for the management and preservation of the site. The acoustic present of Stonehenge is destroyed by the proximity of the nearby roads, and vehicle noise is such that this significant feature of the monument is distorted. In particular it is impossible to appreciate the sound at the Heel Stone much of the time, or to explore the acoustic effects of approaching the site along the avenue and experiencing the changes in sound. These acoustic elements are as much a part of Stonehenge and its landscape as the stones themselves, and as more is discovered about the acoustic significance of the site, its acoustic ecology must become a consideration in plans to rearrange it. This will surely add weight to the voices of those who wish nearby roads to be re-routed or a noise-masking roadside wall to be considered.

There were the prehistoric equivalent of cheap and expensive concert hall seats at Stonehenge, high and low status positions, with measurements of envelopment telling us that being inside the circle indicated clearly a separate level of status. This mirrors the present situation, with only a select few, those who book or are present on special days, allowed within the sarsen circle while the mass of daily visitors follow a path around the outside. It may be that further research into the acoustics of the space will help us to understand why Stonehenge remained a focus of musical activity through the festivals of the 1970s, why there were those willing to fight to hold musical events in the space in the 1980s, and why the Salisbury Festival has in recent years used the space as a concert venue. The sense of envelopment within the space alone provides a sense of involvement that augments the power of communal musical experiences to bond individuals together. It does seem that those who met every year in the 1970s to celebrate the summer solstice with music and dancing may have been echoing the behaviour of their prehistoric ancestors to some extent.

One important research question within this project was to suggest possible descriptions of ritual activities in prehistory. Our experiments showed that circular modes of vibration could be stimulated by loud repeated rhythmic patterns, creating low bass sounds, like a throbbing synthesizer bass or humming. But could this kind of loud bass effect have been created in prehistory? There were few low bass sounds in prehistory except perhaps thunder, earthquakes and the sonic booms of meteorite strikes. (All of these sounds are associated with thunder-god archetypes, providing the intriguing possibility of a link with the symbol of the thunder god, the hammer, carved onto some of the stones at Stonehenge.) Low frequency sounds would have been very unusual and noticeable. In order for these loud sounds to be generated, a large group of people would have to have played at the frequency or tempo associated with that resonance. Echoes in the space may have provided a kind of metronome to keep this rhythmic playing in time. In order to make the infrasonic frequencies audible, the sounds would have to be loud, above 96dB. This is unlikely using human voices alone. However, percussion sounds can achieve this volume, especially considering the loudness of the space caused by its acoustics. It is possible that those able to hear lower than usual (perception of low sounds varies from person to person) would have perceived the low resonances in the space more clearly. Older people can also hear these sounds more clearly, so it may have been the wisdom of the old rather than the strength of the young that characterised participants.

We know that some of the working of stones at Stonehenge was done at the site. Numerous stone chippings have been found, so people were hitting stones and making percussive sounds. We also know that in many cultures, work music and ritual music are often linked, and of course the construction of Stonehenge was itself an important ritual activity. If one plays in time with the echoes in the space, the echoes set a tempo like a metronome. Playing in time to the echoes would be what one would expect: playing against the echo patterns would result in a cacophonous sound. Hitting stones in time with the echoes would have helped to synchronise work. Rhythmic music in the space would have to play in time with the echoes, otherwise it would also become a mess of echoes, whereas playing in time would build up the rhythm, help sound makers to synchronise and play in time together. In addition it would build these powerful bass modal resonances. The space could have started to resonate with modal vibration. The tempo required would be about 156 beats per minute, slower in the winter due to atmospherically driven changes in the speed of sound in air, with quavers played by the music makers, and echoes providing a galloping semiquaver pounding in the centre. We have seen that the acoustics of the space would have been encouraging to rhythmic rather than harmonic music. Dancing to music at this speed would raise the heartbeat and may have helped to create a trance-like effect, overloading the brain's input, enhanced by the high volume of the space and the effects of the infrasonic low frequencies present.

These low frequencies are very similar to the brainwave frequencies of alpha and beta types of activity. It is possible that those around the edge of the stone circle would have chanted at the speed of the percussive sounds, or sung long notes with a vibrato wobbling in time with the music, in order to entrain their brainwaves to the music, to make the dominant frequency of their brain activity slow to a theta pattern, typical of deep meditation, hypnosis or trance. This is similar to the hypnogogic or twilight state between waking and sleeping, and the eyes would be shut for this practice. In the centre a different frequency could have been produced by modal vibration and echoes, one of 10.4Hz, still associated with closed eyes and meditation, and for some with healing. 10.4Hz is associated with alpha wave brain activity, a more active state. It may be that at the centre, a leader was able to stay alert because of this alpha state, continuing to drive a rhythm and communicate with those at the edge of the stone circle as they gradually became lost in a deep reverie. The leader could also communicate with those outside the circle. Turow makes a convincing argument that this kind of auditory driving of the brain's frequencies is possible, and provides evidence from many traditional religious practices:

> More generally, several cognitive psychologists hold that perception, attention, and expectation are all rhythmic processes subject to entrainment . . . In other words, even when a person is only listening to speech or music, their perceptions and expectations will be coordinated by their entrainment to what they hear. Entrainment is fundamental then, not just to coordinate with others, but even to perceive, react to, and enjoy music. Music, as an external oscillator entraining our internal oscillators, has the potential to affect not only our sense of time but also our sense of being in the world. (Turow and Berger: 26–7.)

More information (including a range of scientific and clinical studies) is available on the website of the Stanford Institute for Creativity and the Arts. This is a contentious and complex theory and warrants further study. Brain entrainment is used in various therapies for healing of different types, by therapists as far-ranging as clinical psychologists and New Age shamans.

Flashing lights are often used in combination with pulsing sounds in medical audio-visual entrainment machines. Many people are familiar with the capacity of flickering lights to cause epileptic fits, which can be caused by visual stimuli repeating at 5 or 10Hz, although it is more likely at higher frequencies. It is possible that flickering flames at Stonehenge may have been present, perhaps inside beakers that enhanced audio resonances, or as part of cremations. Certainly flames may have been present if ritual events had been at night. The varying pressure of sound waves can make flames flicker with the same frequency as a sound resonance, appearing to dance and change shape in time with music. In a recording studio I was astonished to find that I was able to make a candle flame jump up and down in time to a 156bpm rhythm, with it also growing and shrinking rhythmically, even at low volumes. A flame in a resonant space can also appear to make musical sound, as well as dancing in time to the music (Sen: 156, 170). Flame usually flickers at between 1 and 20Hz (a fact often used in infrared fire detectors), and when sound is close in frequency to a flame's flicker frequency, the two can become synchronised or entrained. In addition, the powerful waves of changing pressure in infrasonic modal sound may have made patterns in smoke from these flames. This is another area in which my comments are somewhat speculative, and more research is needed. We can say with confidence that any flames present would have moved to some extent in time with any low frequency sound vibrations present. One can only imagine the impact on those present of seeing flames dance in time to music or smoke patterns rising into the sky, illuminated by the strobe-like lights of flames.

Trance may be an entrainment of brainwaves, the settling of the brain into alpha or theta wave activity, as when 'lost in music', losing sense of time as the tempo of the music replaces the normal passage of time. Bringing together sources referring to archaeology at Stonehenge (Parker Pearson *et al* 2006; Parker Pearson 2008) and to trance practices (Becker; Rouget), we can try to imagine what a ritual event at Stonehenge might have sounded like.

On the evening before the midsummer solstice people would have travelled to Stonehenge from Durrington. Walking down the ceremonial avenue, they would have passed a sacred stone as they came to the River Avon, marking the beginning of the labyrinthine journey into the realm of the dead. They would walk along the river to the other ceremonial avenue, this one leading up to the stone circle, the city of the ancestors. At Stonehenge, the ceremonial shamans, priests or ancestor-cult leaders would have arrived earlier. Drummers and percussionists would have already taken their places in front of the inner faces of the stones, where the sounds were loudest. The most senior of them would begin to drum first, standing at the circle's entrance listening for the echoes coming back, and timing his playing to synchronise with the 156bpm rhythms of the ancestral stones. For each quaver beat he played, a semiquaver echo doubled the speed of the simple pulse, and other shorter echoes created flams, like a galloping thunder of hooves. Once his pace had settled, the others joined in, and as the standing wave resonance grew, a low thundering boom would begin to build, like the bellow of a huge bull, or the rumbling of the thunder god's mighty chariot. The sound came from the wheel-like circles of sarsens and horseshoe of bluestones, and it was perhaps accompanied by the blowing of long horns of the same kind as those of cattle slaughtered at Stonehenge. The drumming would have called up the echoes of the ancestors, who would have then interceded with the god of sky and thunder, on the day when the earth met the sun.

As the approaching masses walked through the early morning mists, they would have heard the distant thundering sound. Moving up over the crest of the rise, the sound would become a little clearer and more rhythmic, as they caught their first glimpse of the stone circle. Approaching the huge Heel Stone, the sounds would become louder, until at the stone itself they would hear strange noises and voices, that would seem to fly as if by magic from the centre of the circle. Standing in front of the Heel Stone, they would have heard the low note even louder, too low to sing. In addition a rattling, echoing rhythm could be heard coming from the stones. Walking forward, they would come past the Slaughter Stone and its partner, inside the bank, and suddenly the sounds heard would have changed, becoming louder again.

Now inside the great bank, the stones and the crowds of people would have hidden most of what was happening inside the stone circle, and muffled the resulting sounds. People would have stood inside the bank, unaware of the world outside, drawn in and enveloped by the noises from the centre, drawn into the crowd's dynamic. Arriving groups would mingle with the crowd already there, dancing to the rhythms they heard, and becoming lost in the mass of moving bodies. The flames of torches, lamps or lights in beakers, placed on the stones and wooden posts, would seem to move and dance in time to the music, creating strange patterns of smoke, the flickering adding to a disorientating effect, the darkness and sensory deprivation helping to create altered states of mind.

As time went on, one of the more experienced might have felt a spirit calling him forward. He would have lost track of time, and of himself, his mind drifting into a trance-like state of altered consciousness. Dancing to the fast music would have raised his pulse, and his bloodstream would be full of endorphins, his focus in his body, his mind elsewhere. He would be hoping that one of the ancestors would fill him, taking him another step deeper into their world. As he moved forward towards the stone circle, the sounds from inside would become louder and brighter, seeming to become wider and enveloping him. His feelings would grow stronger, and his excitement and anticipation would also swell.

As he crossed the threshold of the great stone doorway, for a moment the sounds would have become quieter and more muffled, as he passed under the stone lintel. Then suddenly he would be inside, and everything he heard would be far louder and more intense. To his left, right and all around the sarsen circle, ceremonial musicians would be playing simple rhythms, on the bottoms of decorated pots and beakers, on animal skins stretched over circles of wood, horn or clay, or on hollowed-out logs. Some would beat objects with antler picks, others hit stones or sticks together, all together producing a loud fast driving beat at around 156bpm, in time to the echoes that rang around the space. This would generate standing waves, making the rhythms throb like a synthesizer, a low booming note like a sustained bottom note of an organ or bass guitar woven into the rhythm, the sound echoing and ringing in the space. Some of the sounds would seem to come from out of the air, or from the stones themselves, the resonance disguising the direction of sounds' sources as well as softening their attacks.

Looking upward, the trancer would feel as though he were in another world, seeing the sky framed by a circle of lintels. The flickering light would cast strange dancing shadows as he tried to see past stones and bodies. As he stepped forward between the circles of sarsens and bluestones, the sound level would have dropped, and the intensity eased for a moment. From here he would be able to see people in the centre moving their bodies strangely. This would have indicated that they were spirit-possessed, ridden by the spirit of an ancestor, blessed by their presence.

Moving forward between the bluestone ring and the trilithons, perhaps someone would have given him a drink laced with hallucinogenic plants, or perhaps his expectations alone were enough to trigger a trance. Watching those close to him, he would begin to dance, his head nodding backwards and forwards, arms and legs moving in time to the music, the low frequency rumble booming in his head and stomach. As he moved around the centre of the stone circle, sounds would move, shift and change. Drawn in by the excitement, euphoria and collective effervescence of the sound, firelight and shadows, he would lose himself in the world of the ancestors, his consciousness absent, his subconscious free to express itself, as the stones and voices of the dead seemed to come alive, to speak and communicate with and through him, becoming as one with those around him, whether living or dead.

He would not know how long he had been dancing there, seeming simultaneously an age and yet no time at all, his body and mind entrained to the rhythmic time of the ancestors, and enthralled by the strange and powerful feelings evoked. At some point later he might be ushered behind a middle-sized trilithon, where the volume was lower, in order to recover. Later the man would tell his friends of his experience, as his consciousness returned to normal. He would be surprised to find that several hours had passed. He would take the blessings of the ancestors back with him to his home, and his courage in stepping inside the stone circle would be rewarded by the authority that his experiences would give him amongst his people.

Gradually it would grow lighter. If it was an auspicious year, the sun might rise above the horizon in a clear sky, seen above the Heel Stone, shining exactly through the standing stones placed on the perimeter of the stone circle, and directly through the uprights of the largest trilithon. The people would thank their ancestors for bringing life back to the land and making it fertile once more. The young would ask for fertility for themselves, to bring new births, to complete the cycle of the unborn, living and dead.

The ritual marking of the cycle of life and death was just the start of the midsummer celebrations. By the end of that day as the sun was setting, people would have returned to Woodhenge, the great axis mundi or tree of life. The wooden circle was also aligned to the sun, which would now be setting in the west. Celebrations would continue long into the night. The rite of passage that we today call marriage might well have taken place, at the peak of the summer, the peak of the country's fertility. The joining of man and woman would have many cosmological overtones: sky and earth, bull and cow, circle and straight line, phallus and womb, sun and moon, birth and death, wood and stone, day and night. There would have been a great, excessive feast involving drinking, overeating, joyful music and winding circle dancing. The births of children conceived by young couples, who were at these events for the first time, would have also marked the great cycles of rebirth in nature.

At the winter solstice there would again be a ritual event. In the winter the day would start in the morning at Woodhenge. Everything is heading towards death at this time of year, and the land is getting less and less green. The sun itself appears to be dying, and this is that crucial moment when everything is restarted, when the sun is reborn. While the summer solstice was about life and fertility, people returned for the winter solstice to bring their year full circle with a ceremony for the newly dead. On the shortest day of the year thousands would gather inside the wooden circle inside the Durrington settlement. Like Stonehenge, the circle's posts were precisely positioned to frame the rising sun. The wooden circle would be the start of a journey for the dead leading them to Stonehenge. People would have brought the cremated remains of their dead to Durrington, and would have carried them down the avenue towards the River Avon.

Most of the ashen remains would be put into the river so that the dead could travel downstream towards Stonehenge, carried away by the current on their path towards the city of the ancestors. The mourners would follow along the bank of the river, and the day would end at Stonehenge where the spirits of the ancestors would be called to welcome the newly dead to their realm, the rhythms of the drums again calling the voices of the ancestors into the presence of the living, within the stone circle. The rhythms would be a little slower, echoes and the speed of sound slowing in the cold air and lower air pressure. Again some might become spirit-possessed, and commune with the ancestors, seeking their wisdom and assistance, the minds and bodies of the drummers and dancers synchronised to the echoes and rhythms. Just as they were asked in the summer to intercede with the gods of sun and land to bring fertility, they would now be asked to help call back the dying sun, calling again the booming thundering voice of the standing wave resonance.

Some may have come to ask the ancestors why they had been afflicted with illness or bad luck, and what they must do to gain respite, or perhaps to seek the possession of an ancestor for healing of body, mind or spirit. Afraid of entering the space, some may have used its unusual acoustic properties to speak to those within from the Heel or Slaughter Stone. Others would be here for the great hog roast that would follow at Durrington.

The twin sites of Stonehenge and Woodhenge were a set place at which to mark the solstices, the great cycle of life and death, and to commune with the ancestors so that they would intercede with the gods. It seems that music would have played an important role in this ritualised cosmological communication. We may now think of the dead as being under the earth, and gone, but for these people the ancestors were present and active. Music, along with the dancing and spirit possession it inspired, would have been a key ritual technology for this communication.

Obviously this description is just a possibility, but just as postholes in the ground and wall markings are used by artists at archaeological sites to extrapolate an imagined drawing of what a site might look like, sonic markers provide evidence of what might have been experienced in the space, and produce an animated rather than still impression. Some of this is speculative, some is based on the acoustic data and other research. Perhaps this was more like a shamanic rather than a possession trance tradition, where only one person at the centre would go into a trance-like state to meet the ancestors. Perhaps one person at a time was allowed to consult those in the centre, perhaps observers stood on top of the bank. Perhaps no one would have been between the bank and stone circle, although the unusual acoustic effects in this area make this unlikely.

At the very least, the evidence implies that sound and music were involved in ritual activity in the space; that it was fast and rhythmic, possibly at 156bpm with a semiquaver pulse; that flames may have flickered and smoke moved in time to the music; that one person in the centre had the most important position and the highest status when there; that there were participants inside the circle, who became to some extent entranced; that there were others around the edge of the stone circle(s) making rhythmic sounds; that a low booming sound would arise made by pounding rhythms in time with the echoes in the space. Those in the space may have found their brainwaves becoming slowed to an alpha rate of around 10Hz, relaxed but alert, helping them drift into a trance-like state. Some may have had epileptic fits. Extended periods of time in alpha wave brain states may have had physical effects, aided concentration, changed mood, induced visions, and effected healing activities.

Measurements of envelopment, and levels of other acoustic effects, tell us that the centre of the circle was the most important position at Stonehenge, as one might expect, and shows a hierarchy of participation, being within the stone circle the next level of significance, within the bank a significant step less important. It will be interesting to see if earlier phases have less stratification. The Heel Stone seemed to be a significant position for people to stand. The level of envelopment made participants feel very much involved and helped participants bond and achieve perceived alterations of consciousness. The Stonehenge circle was an 'other' acoustic space, different from other natural or man-made spaces, the acoustic adding to its special status. The acoustics make it a likely place for people participating in inclusive ritual activities, a place perhaps for spending time and communicating with the ancestors and gods rather than leaving them to rest undisturbed. The Slaughter, Station, Altar, Portal and Heel Stones seemed to cause echoes in the space, and this may be part of the reason for their exact placement.

More research will further establish details of acoustics in the space. This may involve further analysis of digital models, for example investigating the Altar Stone upright or not and adding a partner for the Heel Stone and the Slaughter Stone. It will involve studying further the differences of acoustics in various phases of the monument's development in order to understand what we can learn from how the acoustics of the space changed over time. The acoustics of the wooden circles at Durrington Walls and Woodhenge can be compared with those at Stonehenge, as can similarities with and differences from other stone circles. The results of this study need to be discussed in detail with archaeologists, ethnomusicologists and others. Eventually it is hoped to create an interactive digital 3D model of the site that will feature sound changing as one navigates around it, perhaps also allowing travel through time to hear how the site develops. Meanwhile, more detailed information about the acoustic properties of sarsen and bluestones, improved accuracy of the digital model analysed by Odeon, and other work should improve the certainty, specificity and reliability of results. Recent acoustic field tests at Stonehenge itself have provided evidence remaining today that has helped to confirm some of these ideas, and these results need to be explored more fully.

I have deliberately reached out beyond absolute proof, into the world of possibility, particularly when describing musically based rituals, in a way that some will find is extrapolating too much from the evidence, and perhaps this is a little provocative. I hope that it will encourage others to consider how much of what I suggest is likely to be correct, and what other corroborating evidence there might be. It is perhaps a useful process for the creation of a hypothesis that can be explored through further study. Interpretation always requires a level of storytelling, and our imaginations are a useful tool in exploring possibilities and likelihoods.

This project in no way claims to have decoded Stonehenge or explained its meaning and purpose. It does not suggest that Stonehenge was created as an outdoor concert or dance hall, music venue or amplifier. It does suggest that music, alongside other visual, astrological, ritual or cultural elements, would have had an important part to play, and that this is worthy of investigation. It hopes to show that music, sounds and acoustics were likely to have been an important part of this iconic site, and perhaps a key element in trance or spirit possession rituals for communication with the ancestors (Parker Pearson *et al* 2006). By doing so it demonstrates the kind of information that it may be possible to discover about archaeological sites through a careful and detailed study of their acoustics, using a methodology that includes contextualisation, background research, theoretical analysis, the use of digital and scale (or full-size) models, and on-site field measurements.

The songs of the stones at Stonehenge have been heard in prehistory, in the era of Thomas Hardy, and at the rock concerts of the 1970s; they can still be heard today. When visitors enter the Stonehenge site, perhaps as well as looking around they will be encouraged by this work to begin to listen, and thus more fully understand one of the most mysterious archaeological sites in Britain.

ACKNOWLEDGEMENTS
I wish to thank Dr Ben Chan, Dr Roger Doonan, Prof Michael Parker Pearson, Dr Graham McElearney, Dr Christina Tsoraki, Andrew Lines, Dr Bill Bevan, Prof Chris Scarre and Prof Jian Kang for their help, advice and support with this project. Dr Fazenda contributed substantially, providing acoustic expertise and analysis of results from Maryhill, carrying out acoustic field tests, and discussing and developing the project with me.

PART II

CULTURAL HISTORY

CHAPTER 5

The cultural history of Stonehenge

Ronald Hutton

Stonehenge has a place in the history of culture that is one of the longest, if not the longest, of any megalithic monument in the world, commencing well before such structures generally began to attract attention: it has been the object of steadily recorded speculation and controversy for almost 900 years.

Over that great span of time, five themes have repeatedly characterised its place in the cultural imagination. The first is its uniqueness: it simply does not look like any other human structure. This is because it represented a one-off experiment, hugely ambitious and daring, in treating stone as if it were wood. It is in fact a woodworkers' wonder, the creation of a bunch of megalomaniac carpenters who shaped and polished off long blocks of stone and fitted them together into uprights and lintels, using mortise and tenon joints. This was and is one of the enduring techniques by which timber buildings are constructed, but only here was it applied to megaliths. The result is the series of three-piece settings of great stones, which give Stonehenge a distinctive logo, a visual motif instantly recognisable as belonging to no other structure.

Its second enduring feature is accessibility: that it stands in the heartland of one of the world's leading nations: the one, indeed, which was until recently the wealthiest and most militarily powerful of all, and which gave the planet its dominant language. Until the 19th century, it is true, the monument was set on the edge of the vast uninhabited grazing lands of Salisbury Plain. None the less, a major thoroughfare ran down the Avon valley to the east, passing through the market town of Amesbury only a few miles from the stones. The site therefore gave visitors the best of both worlds, requiring some effort to reach and enabling authors to enthuse about the awe-inspiring desolation of nature all around, and yet easily attained on foot, let alone on horseback or by carriage, from the nearest urban centre. In the 1840s a new highway was built from London to Exeter which ran right by it (and still does, as the A303), swiftly followed by the appearance of a railway station at Amesbury. The invention of the bicycle and motor car before the end of the century further reduced the difficulties of a visit, either to break a journey or as a day trip from the capital or provincial towns. Around 1870 local people, leavened from the start by Londoners, developed the custom of gathering at the monument each summer solstice to witness the sunrise on which its ceremonial entrance was aligned. Within two decades the crowds that did so had swelled to thousands, and this remains the longest modern tradition associated with the site (Worthington 2004: 14–21; Hutton 2009: 348–9). The development of the Plain as a military training area during the First World War surrounded Stonehenge with army and air force bases, and cafes were built nearby for tourists. At the present day, the great problem of its situation within the human landscape is one of congestion and overcrowding: successive official plans have been drawn up to close or re-route the surrounding highways.

The third key feature of the monument is mystery. The first person to write a scholarly appraisal of it was William Camden, in the Elizabethan period, and he emphasised the fact that nobody could say with certainty who had erected it, and why (251–4). In 1695 an enlarged edition of Camden's book was produced by a team of Oxford academics led by Edmund Gibson. All that changed when it came to revise the entry on Stonehenge (107–10) was that the list of plausible hypotheses regarding its date and significance had become slightly longer and more elaborate. By the Victorian period both the majority of the scholarly works which dealt with it and most of the growing number of tourist guides which were published upon it took the same line: to repeat or volunteer different explanations for the structure while emphasising that ultimately nobody could prove any of them (in chronological order: Hoare; Eason; Zillwood; Michael; James; Anon 1884 and 1894; Goddard). There were some overall advances in knowledge and certainty, the most important being that by the 18th century a consensus had emerged that Stonehenge was constructed by the pre-Roman British, while during the 20th better dating techniques placed it with increasing firmness in the late Neolithic period. It was also definitely accepted as some kind of ceremonial structure (Chippindale 3/2004). None the less, most of the scholarly literature upon it has continued with equal consistency to emphasise our ongoing ignorance of the purpose for which it was erected, the circumstances under which this occurred, and the religious rites and beliefs of its builders (for example, Bradley: 138–42; Cleal *et al*; Hill; J Richards). This uncertainty produces a large part of the appeal of the monument to the general public: the imagination of ordinary people can play freely upon it, largely undirected and unrestricted by expert opinion.

This general confession, and celebration, of ignorance permits the fourth theme in the cultural history of Stonehenge: the breakthrough hypothesis. This is a theory which claims to disperse the mystery by providing a clear and plausible explanation for the purpose of the monument and the circumstances of its construction; and which wins many adherents. The two most successful to date have been those of Geoffrey of Monmouth, which was repeated as fact by many writers for half a millennium, and of William Stukeley, which was dominant in interpretations of the site for two centuries (Chippindale 3/2004: 1–156). Geoffrey, publishing around 1140, declared that it had been a war memorial designed by the wizard Merlin, using the supernatural powers of wisdom inherent in being the son of a demon. This had the virtue of drawing attention to the unique significance of the stones, and also of inviting scholars to try to relate it to particular human cultures. Stukeley's theory, which appeared in 1740, was that it had been a temple of the British Druids. This also had virtues, notably in persuading almost all people that it was the work of pre-Roman natives and in drawing their attention to the landscape in which it is situated and the many other prehistoric structures that feature in it. As his ideas finally lost their hold on the popular imagination in the 1960s, they were replaced by those of Gerald Hawkins, who declared that it had been an ancient computer, used among other purposes to predict eclipses, and Alexander Thom, who argued that it had been an observatory, aligned upon various different heavenly bodies. Suited to the contemporary cult of science and technology, these twin hypotheses had their own virtues, of compelling prehistorians to look more closely at the mathematical and astronomical associations of the monument.

The present time is unusual, in having produced two competing explanatory frameworks, suggested simultaneously by groups of archaeologists of equal reputation, based upon equally impressive programmes of research, and receiving equally colourful and enthusiastic presentation in the mass media. Proponents of both are, indeed, represented in this collection of papers. The first is championed by Tim Darvill and Geoffrey Wainwright: that Stonehenge was essentially a temple of the living, dedicated especially to healing rituals to which people came on pilgrimage from long distances away—a kind of prehistoric Lourdes (Catling; Darvill 2006). The second theory is advanced by the Stonehenge Riverside Project, most prominently represented by Mike Parker Pearson and including Joshua Pollard among his many colleagues. It conceives of the monument as primarily a place of the dead, visited by the living at numinous times of year to honour their ancestors and the recently deceased members of their community; most notably at the winter solstice, after enjoying pig roasts at the huge nearby enclosure of Durrington Walls (Parker Pearson 2007; Pitts 2008). Both interpretations are supported by an adequate body of material evidence, neither seems entirely susceptible of proof, and they may not indeed be incompatible. To raise the stakes, however, they have also each come up with a sequence of construction for Stonehenge, based on new dates, which directly contradicts the other (Selkirk). It remains to be seen how long they stay in balance, and what succeeds them; the one certain conclusion is that, in their focus on community and on religion, and in their parallel reconstruction of essentially peaceful and co-operative Neolithic societies, at harmony both internally and with their neighbours, they reflect the preoccupations of the present time as clearly as the linkage of the stones with big science did those of the 1960s.

The fifth and last theme in the cultural conception of Stonehenge, and the one which has the most direct bearing on the way in which it has 'sounded', is that of cultural prejudice. Because so much uncertainty is invested in any consideration of its original nature and purpose, members of each successive generation can infuse their views of it with their own attitudes to past and present humanity. These attitudes certainly change collectively over time, as has been shown above, but there are also enduring aspects to them which recur century by century. Two in particular have coloured the cultural reception of the monument ever since discussion of it began in the Tudor period. Stuart Piggott characterised them neatly as the 'soft' and 'hard' views of primitive societies: they might be extended, as Piggott's own writings starkly illustrate, to views of humanity as a whole (2/1975: 91–3). The 'soft' view tends instinctually to credit older and simpler societies to one's own with a greater inherent goodness and wisdom. This is rooted in the Greek and Roman classics, where human society was held by some writers to have degenerated from an original golden age of peace, kindness and equality. Some of these authors held that remnants of it had been preserved most completely among those people who continued to live at the simplest level of economy, society and technology. Christianity added a further dimension to this view by teaching that all humanity originally practised the good religion revealed by the single true God, once in the time of the early descendants of Adam, and once more in that of Noah and his children. Evil was due, in this view, to the degeneration of religion, and therefore of general behaviour, which set in subsequently, as a result of fallen human nature. This viewpoint usually held out the hope of redemption, by a return to the purity of the original faith and the simplicity of the earliest society.

The 'noble savage' view of the remote past and of traditional peoples has always had a particular appeal to people who dislike aspects of their own society, or at least its present form. Since ancient times a contrast between noble simplicity and corrupt sophistication has been used to castigate highly-developed civilisations from within; in that sense it is one natural manifestation of a counter-culture. In its Biblical form, however, it has also had great purchase on mainstream Christian culture: in the 1840s the story of the human race as one of degeneration from primeval goodness and truth was inscribed on the pediment of the British Museum, where it remains. During the Georgian period it was dominant in European culture, as intellectuals sought to draw together the different branches of humanity into a common family, and their differing faiths into a common framework. This is the vision that would people Stonehenge with sacred scientists, natural environmentalists, pilgrims seeking healing, and/or peaceful worshippers seeking communion with ancestors. It lies behind all the recent 'breakthrough' interpretations of the site mentioned above (Hutton 2007: 41–78).

The countervailing, 'hard' view of prehistory instinctually associates the primitive with savagery, ignorance, dirt and superstition, and characterises primeval religion as one of gloom and gore, exemplified in the practice of human sacrifice. It is also rooted in antiquity, being deployed by most ancient civilisations, but above all the Romans, against tribal peoples beyond their borders. Equally, it may be found in the Bible, used by the ancient Hebrews to condemn the polytheist cultures that surrounded them in the name of the cult of Yahweh. This negative language received a new impetus with the onset of the Victorian period, for two reasons. One was the rapid expansion of European colonial empires, of which the British was the largest, into territories populated by traditional peoples in other continents. The greatest moral justification for this process was to civilise the natives of these lands, and so improve the general condition of humanity: which required a proportionate emphasis on the backwardness and inhumanity of the cultures concerned. This attitude was back-projected onto the ancient European past by the second relevant feature of the period, a cult of technological progress coupled with Darwin's new theory of evolution. This led to the condemnation of uncivilised societies, past and present, as inherently brutal and deplorable. The union of the two concepts was painted onto a corridor of the newly rebuilt Houses of Parliament in the 1840s, in a pairing of pictures devoted to 'the Progress of Britain'. The first showed an ancient British Druid committing a human sacrifice, the second the contemporary British campaign to suppress the religious custom of widow-burning as part of the conquest of India. The message was clearly that the British had escaped the barbarism of ancient times, and so become fitted to eradicate that remaining in the present (Hutton 2007: 93–136).

It was this attitude which, for the past half millennium, has regularly populated Stonehenge with human victims pouring out their life-blood on reeking megaliths or burning alive in wicker cages. It has been written onto the site by the use of the name 'Slaughter Stone' for the recumbent megalith inside the main entrance of the surrounding bank, coined by a religious fanatic called Edward King in 1799 (159–209). King's reasoning in bestowing it was based upon a series of outright factual errors; but it has stuck ever since. This vision of the monument as a setting for scenes of Gothic horror has dominated literary portraits of it, whether traditional, like those of William Wordsworth or Thomas Hardy, or recent, in the novels of Henry Treece (1956), Edward Rutherfurd (1987) or Bernard Cornwell (1999). It has a clear potential in providing titillation or drama, but also a more serious role in the politics of religious intolerance. Repeatedly over the past 500 years the 'hard' view of prehistoric religion has been used to condemn, by association, aspects of belief against which authors have had an especial animosity: the list of targets includes Roman Catholicism, paganism, established churches, atheism and (following the events of 9/11) Islamic fundamentalism.

This, then, is the peculiar strength of Stonehenge as cultural icon: it is an instantly recognisable structure which resembles no other, and onto which a remarkable range of fantasies can be credibly projected. It is undoubtedly and enduringly *useful*.

CHAPTER 6

Megaliths in English art music

Stephen Banfield

Meaning in music

What music means is a perennial problem for its role in the academy. Peter Borsay, trying to explain why historians have paid so little attention to music, puts it rather well (98–102). He identifies three musical footholds the historian may try to gain. First, there is the social meaning of musical exchange as commodity, ritual or performance. Second, there is the meaning of the words that accompany vocal music, the action that stage music or ritual music accompanies, or the title that heads the work. So far so good. The third is much more slippery: the meaning of the notes themselves. Despite a classicised 18th-century aesthetic whose conventional wisdom was that music could 'only' represent objects from another art or sensory sphere, and despite its transformation into a romanticised 19th-century one in which music's meanings were felt to be far deeper than 'mere' pictorial or sonic representation, it remains highly problematic to say that a piece of art music is in any sense 'about' a megalith; yet that is what I am going to write about.

At one end of the spectrum, Thomas Lloyd Fowle's 'The Stonehenge polka' for piano, published in 1855, is in no sense musically 'about' Stonehenge except as the usual metonymic displacement for its sheet music cover, a rather charming lithograph of the monument which may or may not be of interest to archaeologists and cultural historians (Fig. 6.1). The sounds themselves are purely generic, comprising a utilitarian piece of dance music like thousands of others, although the attempt to turn 'God save the Queen' into eight-bar phrases (why refer to this tune at all?) in the final section adds some curiosity value (Fig. 6.2). Fowle (1828–96), whose father was the longest-serving rector of Amesbury, operated in Winchester and typifies the freelance genteel musician of the time, his professional dignity, itself defensive (see Dibble: 172–4; Horton: 249), belied by the banality of his music. At the opposite end, I want to concentrate on the orchestral and piano works of John Ireland (1879–1962), who seems to me to offer a profoundly satisfying subjectivity of relationship between a modern listener and an ancient monument.

A survey of the landscape

Evidence of megaliths in English art music is not on the face of it very plentiful. By some uncanny coincidence, Fowle of 'The Stonehenge polka' is joined on the list of megalithic composers by Ernest Fowles, no 2 of whose *Five Reminiscences* published in 1919 is entitled 'The Rollright Stones'. Before this, the latest manifestation of a social dancing craze had produced two novelty piano pieces, of 1904 and 1913, probably British rather than American in origin (though it is difficult to be certain), both with a cover illustration by W George which appears to show the same couple a decade apart in time, Luke Everett's 'The "Prehistoric": cake walk and two-step' (1904) and Norman Kennedy's 'Prehistoric zig-zags: intermezzo 2-step' (1913) (Figs. 6.3 and 6.4). In an extraordinary contrast, the first of John Ireland's three single-movement orchestral works evoking prehistoric sites and pagan rituals also dates from 1913. This is *The Forgotten Rite*, associated with prehistoric Jersey; the other two works are *Mai-Dun* of 1921, referring to the largest English earthwork, Maiden Castle in Dorset, and *Legend* of 1933, with concertante piano, creatively connected with Wepham Down and Harrow Hill on the South Downs. The year after *Legend* was composed, the first opera of George Lloyd (1913–98), *Iernin*, was premiered in Penzance, only four miles from the Nine Maidens stone circle of Boskednan where its first and third acts take place (in the 10th century). Ireland later wrote a piano suite, *Sarnia*, celebrating Guernsey, where he lived in 1940; its first movement, 'Le Catioroc', depicts pagan rituals at *Le Trépied* dolmen on the island's west coast. After this, megaliths in English music appear to have had a lean time until more recent years. Vaughan Williams's 9th Symphony of 1958, his last major work, owed its gestation to a conceptual narrative concerning Salisbury Plain whose traces survive in the second movement insofar as it is known to refer to the capture of Hardy's Tess at night-time Stonehenge and her subsequent execution by hanging. But it was only in the 1970s and 80s that a much younger composer was to be found writing music about megaliths and associated prehistoric sites, namely Harrison Birtwistle in *Silbury Air* for chamber ensemble (1977), which refers to Silbury Hill, and his opera *Yan Tan Tethera* (1984), which takes place among the sarsen stones of the Marlborough Downs. Curiously for one who has chosen to live on Hoy in the Orkneys, close to some of the most impressive megaliths and prehistoric burial chambers in northern Europe, Peter Maxwell Davies seems scarcely to have referred to them in his music. But my own colleague John Pickard has composed what may be the most overt and comprehensive megalithic treatment yet in music, *Men of Stone* for brass band (1995). This symphonic suite consists of four continuous musical sections identified as 'Avebury (Autumn, morning)', 'Castlerigg (Winter, afternoon)', 'Barclodiad-y-Gawres (Spring, evening)' and 'Stonehenge (Summer, night/dawn)' and was later incorporated into Pickard's *Gaia Symphony* for brass

FIGURE 6.1: THOMAS LLOYD FOWLE, 'THE STONEHENGE POLKA', SHEET MUSIC COVER

FIGURE 6.2: FOWLE, 'THE STONEHENGE POLKA', SECTION INCORPORATING 'GOD SAVE THE QUEEN'

FIGURE 6.3: LUKE CAVENDISH EVERETT, 'THE "PREHISTORIC"', SHEET MUSIC COVER

FIGURE 6.4: NORMAN KENNEDY, 'PREHISTORIC ZIG-ZAGS', SHEET MUSIC COVER

band as its final movement. Curiously, another Bristol colleague, Geoff Poole, has also composed a 'stone' piece: *Carved in Stone*, commissioned by the Corsham Festival in 2006 to celebrate the southwest landscape (he lives in Bradford-on-Avon). The last of its three movements is entitled 'Avebury'. The remaining megalithic music I have managed to locate is very much peripheral to this chapter. David Little's *Stonehenge Studies* (nos 11 and 12 in the sequence published in 1992) come from an American; Gareth Wood's *The Margam Stones* for brass band (1979) refers to early Christian inscribed stones in Wales; Antoine Tisné's *Music for Stonehenge* for alto saxophone and piano (1980) is French; and *Prehistoric Piano Time* (1996) by Pauline Hall and Paul Drayton is a children's piano book of '20 prehistoric pieces and puzzles' cashing in on the dinosaur craze (one of the puzzles has the notes of the musical scale hidden among the scales of a 'scaleosaurus'). Paul McCartney's orchestral piece *Standing Stone* (1997), based on his own poem, continues his bid for classical status begun with *The Liverpool Oratorio*. Finally, 'An Avebury carol' by Robin Nelson (2003) should not be overlooked.

Types of cultural transaction

It would seem that there are four types of cultural transaction taking place in these works, operating chronologically. The first is that of the sentimental or wry souvenir. The second, third and fourth embody geopolitical contrast and indeed conflict, between the hegemonic suavity of a southern English pastoral, the wild independence of the Celtic renaissance and the hardness of northern devolution.

As a souvenir, Fowle's 'The Stonehenge polka', at least its music, fosters a mixture of irreverence and pride—the jauntiness of a hoary monument preparing to meet the latest youth craze, rather as the mateyness of bumper stickers a century later would pass on to the occupants of the car behind a newly jollified notion of a day out at a stately home with 'We have seen the lions of Longleat'. There are better-known examples of 19th-century musical souvenirs, including Johann Strauss senior's dance music with titles such as 'Eisenbahn-Lust-Walzer' ('The railway waltzes'), and others that, like Fowle's piece, take a fashionable delight in squeezing earnest tunes into party costume, notably the *Souvenirs de Bayreuth* and *Souvenirs de Munich* concocted by Fauré, Messager and Chabrier to shoehorn Wagner's leitmotifs into quadrille dance rhythms. But the cakewalk and two-step take the irreverence several stages further, their musical analogue to George's cartoon garishness being the syncopation which in the early 20th century threatened everything old with mockery and, by implication, destruction.

Ireland, Fowles with his Rollright Stones, and Vaughan Williams exploit the cultural capital of the southern pastoral (see Howkins, and chapter 7 below). Ireland's and Vaughan Williams's prehistoric orchestral scenes are set in the most easily accessible geographical area of the south that has both wide open spaces and plentiful evidence of primeval habitation, the chalk downlands, and the best articulation of the sensibility that gravitates towards them comes from E M Forster, who in his short story 'The curate's friend' wrote: 'It is uncertain how the Faun came to be in Wiltshire . . . [but] any country which has beech clumps and sloping grass and very clear streams may reasonably produce him' (90)—and who, you will remember, produced the character of Mr Beebe in *A Room with a View*, affording us our moment of vicarious paganism watching Simon Callow, Daniel Day-Lewis and Julian Sands frolicking naked in some leafy hammer pond of deepest Surrey. The keynote here is lost or cherished spirituality: the need to retreat to the rural landscape, and to some notion of it as an inheritance, as counterpart to the pressures of urban modernity. Actually it represents the bourgeois citizen's privileged opportunity to do so, for it is impossible to dissociate the artistic urge from the economic ability to fulfil it. Ireland could afford lengthy working holidays on Jersey and Guernsey and, eventually, to buy a windmill where he lived in Sussex with a view of Chanctonbury Ring. While formulating the Ninth Symphony, Vaughan Williams revisited in his mind and indeed in his car the Wessex of youthful walks and youthful tone poems, some of whose music he would incorporate into the new work. Photographs taken by Ireland while revisiting Dorset with a handsome young friend two years after completing *Mai-Dun* (Fig. 6.5) testify to the monument's ongoing touristic appeal to the composer, to which in this case an erotic charge adds capital, while rendering it literally picturesque. To what extent this romantic urge is a 20th-century continuation of primal festive compulsions might well be asked.

Fowles's *Five Reminiscences* are a slightly different commemorative case, being dedicated 'To my Son E L Douglas Fowles. (Tank Corps: Bohain. October 11th 1918)'. These piano pieces, published in London by Winthrop Rogers, are entitled 'Dewberries', 'At the Rollright Stones', 'Between whiles', 'Field-flowers', and 'Compton Winyates'. The north Oxfordshire locations of two of them and rural keywords attached to two others presumably enshrine moments in places shared by father and son, dead, absent or wounded as he must be, whether as permanent or holiday residence.

The third type of megalithic trope, the Celtic, is exactly opposed to this and celebrates hard-won freedom from the subduing cultures of Anglo-Saxon hegemony. Since my topic is *English* music, this more or less limits us to Cornwall, and even to one work, for although both Arnold Bax in his orchestral tone poem *Tintagel* (1919) and Ethel Smyth in her opera *The Wreckers* (1906) were inspired by monolithic stone structures, namely a castle and a cave (Piper's Hole, the sea cave on Tresco in the Isles of Scilly), neither invokes a megalith. Lloyd's opera *Iernin*, by contrast, does so in no uncertain terms (Fig. 6.6). As mentioned earlier, Acts I and III take place at the Boskednan stone circle. Iernin, the faery heroine, was along with her sisters turned to stone by an early saint for making love to mortal men but now at the beginning of the opera resumes human shape. The hero Gerent is supposed to be marrying the beleaguered Prince Bedwyr's daughter Cunaide in order to keep the Saxons at bay (they have their designs on her), but on seeing Iernin he is, as the sleeve note puts it, 'completely

FIGURE 6.5: PHOTOGRAPHS OF MAIDEN CASTLE AND ARTHUR G MILLER TAKEN BY JOHN IRELAND, 1923

FIGURE 6.6: GEORGE LLOYD, *IERNIN*, POSTER (1935)

FIGURE 6.7: JOHN PICKARD, *MEN OF STONE*, FRONT COVER

FIGURE 6.8: HARRISON BIRTWISTLE AND TONY HARRISON, *YAN TAN TETHERA*, OPERA FACTORY PRODUCTION. 'BAD'UN' AND STANDING STONE IN THE BACKGROUND

glamoured'. After a night of passion, Gerent is persuaded by Cunaide to resume his political duty, and Iernin returns to her place in the stone circle.

Lloyd's opera is extraordinary in many ways, not least because here was a 21-year-old upstaging senior composers such as Vaughan Williams who, as he put it to the youngster when *Iernin* enjoyed a nightly run at the London Lyceum in 1935, had been trying unsuccessfully all his life to write opera. As recounted by Lloyd (*Iernin*: interview), *Iernin*'s luck began when Frank Howes, the *Times* music critic, caught its premiere while on holiday in the Penwith peninsula in November 1934, and no doubt it was a walking holiday, for as he wrote in his review: 'To anyone who knows the part and has floundered through bogs and gorse to these unapproachable stone maidens it is as queer an experience to listen to the translation of their history into music in Cornwall itself as to hear *The Mastersingers* in the city of Nuremberg'. He liked the work, adding that 'the vividness of local colour and atmosphere is no sufficient testimony to the power of this new opera' (*The Times*, 7 November 1934: 14). Its power resides in the composer's refusal, in contrast to Bedwyr's subjection, to come under the jurisdiction of English musical Celticism of the Rutland Boughton variety: Lloyd specifically mentions this disinclination in his interview with Chris de Souza. In a duchy still at that time believing in a Phoenician heritage, and with a passionate lover of Italian opera at his elbow in the shape of his father, the opera's librettist, he turns to the Mediterranean for his style and dramatic stance. The result for the megaliths is modal Verdi—rather as though a gecko were basking on lichen. The moment at the end of Act I where Iernin is left alone after her initial encounter with Gerent demonstrates the point (CD1, track 6, 4'59"–end).

The more recent megalithic music enumerated above, our fourth type, tends to promote northern toughness rather than the softness of western Celtic romanticism, but may favour a Celtic alliance nonetheless, if the front cover of the score of Pickard's *Men of Stone* is anything to go by (Fig. 6.7). Here we see higher hills than any in the south, men rather than a woman of stone, runic writing for a 'bardic' edition, and the Borders-sounding name 'Kirklees', establishing the aura of masculinity one would expect for a brass band score and fulfilled by the music itself, percussive, immediate in its dynamic expression, full of no-nonsense rhythms and never sentimental in its reflective passages. Poole's programme note for the first performance of *Carved in Stone* contrasts the northern landscape in which he used to live with that of his new southern home. But it also seems more responsive to tropes of femininity than masculinity when it explains how 'Avebury''s musical structure corresponds to a cyclical idea of 'nine Beltane dawns from across the centuries' and its affect to seeing the sarsen stones as a 'theatre of goddesses'.

Northernness in Birtwistle's *Yan Tan Tethera* is a rather different matter from Pickard's, for as an opera it returns to Lloyd's method of dealing with geopolitical conflict by making a story out of it. In setting Tony Harrison's libretto to music, Birtwistle manages to celebrate the ancient mysteries of the Wiltshire downs (where, incidentally, he now lives) without capitulating to their southern values by creating as his hero a 'good' northern shepherd, Alan, who has travelled south but is missing his home country. Singing, as Omar Ebrahim did in the televised Opera Factory production, in a flat northern accent, he affirms in his opening aria, 'I came from t'North with one ewe and one ram and I don't care much for this place where I am . . . I long to go back . . . away from the chalk downs and towering stones—back to rough crags and the Northern fells where the rocks are steep and the moor winds moan and sheep don't go round waving bells . . .' The black-faced southern sheep (there are two flocks on stage) are by implication effeminate, and his adversary, Caleb the bad southern shepherd, admits to having stolen and melted down the local church bell for his flock's tintinnabulatory embellishments. Furthermore, aided and abetted by the 'Bad'un' in the form of a piper, he is stealing gold from the prehistoric mounds on which he sits while the sheep graze and will steal Alan's southern wife Hannah. Much of the opera's cogency, as throughout Birtwistle's output, stems from the dramatic and musical use of ritual repetition or circulation (the word 'ritual' appears 19 times in Jonathan Cross's *Grove* article on the composer). This forms a satisfying analogue to, indeed interpretation of, the ancient stone circles, and manifests itself most fully in the libretto in sequences of counting, focused naturally enough on counting sheep. Counting to three is the meaning of the opera's title in one of the old northern number systems (the opera actually uses the Lincolnshire version), and it forms the beginning of the northern charm Alan teaches Hannah: 'Yan Tan Tethera 1 2 3 Sweet Trinity Keep us and our sheep.' The three uprights of the stone dolmen in the middle of the stage are also trinitarian, though there are plenty of twos in the opera as well—two shepherds, two cycles of action, two sets of twins, and in the Opera Factory production an effective doubling of piper and anthropomorphic standing stone (Fig. 6.7). Not without humour, at the end of the opera Caleb is defeated by his own misuse of the spell and shut up in the dolmen, whence his eternal cry of 'Tethera dik'—thirteen—can sometimes be heard.

John Ireland

But to return to John Ireland. His one explicitly megalithic piece, 'Le Catioroc', would at first sight seem to confirm an indulgent aesthetic of the romantic picturesque. The first of three movements in a suite published in 1941 as reminiscences of an idyllic period spent on Guernsey, it is suffixed 'Fort Sausmarez, L'Erée, 1940'—written on the spot, therefore, with the purported authenticity of a *plein air* painter. Given that the second movement of *Sarnia*, 'In a May morning', is a sentimental reminiscence of a boy with whom Ireland was briefly infatuated in St Peter Port in May 1940, and that the third, 'Song of the springtides', evokes the seaspray of the island at a heady time of year, one might write off *Sarnia*, even more than *Mai-Dun*, as holiday snapshots posted in a nostalgic blog. This was a highly charged blog, to be sure, for Ireland, showing every sign of wanting to settle in Guernsey, so happy was he there, was forced to leave a month later on one of the many crowded boats frantically carrying evacuees back to the UK only a week before the islands were given up to the Nazis. But that is not the whole story. Ireland and a friend shared a rented house at Fort Saumarez, on a headland of the wild west coast of Guernsey. When they first saw it, Ireland said to his companion, 'John, we are in one of those strange places' without knowing that on the hill immediately behind was another of Guernsey's dolmens, *Le Creux ès Faies*. And while they were there, a singer staying with them, Maria Scottie, had a horrific vampire-like experience one night that resulted in two red marks on her throat in the morning. Something similar had happened to an unconnected woman around the same time in Fort Saumarez, it transpired two weeks later. With the one dolmen associated with fairy lore and the other, on the next headland, supposedly the venue for a witches' sabbath in the 16th and 17th centuries (the last survival of pagan practice on the island), there was bound to be more to Ireland's piece than musical impressionism (BBC Guernsey; Longmire: 46, 56–8, 62, 69–71; F Richards: 72).

That 'more' would appear to be a complete sense of identification with the pagan ritual. The score carries an epigraph from *De Situ Orbis* by Pomponius Mela, a Roman geographer of the 1st century AD: 'All day long, heavy silence broods, and a certain hidden terror lurks there. But at nightfall gleams the light of fires; the chorus of Ægipans resounds on every side: the shrilling of flutes and the clash of cymbals re-echo by the waste shores of the sea'. A plaintive shepherd's-pipe arabesque and throbbing drone convey the initial pastoral calm, but the arabesque appears to invoke more than it bargained for as it becomes increasingly florid. While inevitably also reminiscent of the textures and gestures of Debussy's *L'Isle joyeuse*, a grotesque central section represents the abandon of the orgy by means of jazz techniques that to a classical musician of Ireland's period and persuasion were considered obscene: determinedly swung keyboard rhythms, rapidly pounding boogie-woogie chord repetitions, lasciviously phrased licks and blue subdominant 7ths in tritonal substitution (Fig. 6.9).

Within a ternary musical form, the pagan forces climax and scatter, leaving behind only the plaintive

FIGURE 6.9: JOHN IRELAND, *SARNIA*, FIRST MOVEMENT ('LE CATIOROC'), EXTRACT

pastoral arabesque and drone; but Fiona Richards, in her book on Ireland, cites Tzevetan Todorov's 'ideal narrative' at this point: 'a stable situation ... is disturbed by some power or force. There results a state of disequilibrium; by the action of a force directed in the opposite direction, the equilibrium is re-established; the second equilibrium is similar to the first, but the two are never identical' (Richards: 74). Indeed they are not identical: nothing is quite the same again; something fundamental has been revealed—or disturbed.

This leads us back to a particularly potent form of Borsay's difficult third question: can there possibly be anything in the notes themselves that accomplishes this? Can music do something to or for the stones that other representational art forms, even the humanities cannot?

Perhaps composers can do two unique things. First, they can suggest the sounds of prehistory themselves, which takes us back to the title of this collection. Second, they can use those sounds to depict and even to operate some system of resonant efficacy—to show us the enchantment of others or to enchant *us* into a field of communication, the meaning of the word 'enchantment' of course implying singing.

Yan Tan Tethera's musical charm is given in Fig. 6.10 a. It has certain things in common with the main themes from Ireland's *Legend* and *Mai-Dun* (Fig. 6.10 b and c). One is the provision of a triplet figure at the end. This is the most obvious way of signifying the call of a horn. The very fact that we refer to a horn 'call' indicates its traditional association with long-distance communication, be it for the hunt, the post, an Alpine herd or, here, a being on another plane of existence—a spirit. The opening of Ireland's *Legend* is pure horn call, in this case of the *ranz des vaches* type, but so is the tail end of the opening of *Mai-Dun*, recognisably similar in shape to that of the *Yan Tan Tethera* spell. An equally important common feature is what music analysts call 'developing variation', a doctrinaire procedure for many art-music composers after Brahms, to whose music the term was first applied. Here the term simply indicates the one-two-three-ness of the themes, obligatory for Birtwistle because that is exactly what the words 'yan tan tethera' mean. Something is stated, it is stated again in a slightly different way, and when it is stated a third time something more happens—a door opens, we move off on to another plane.

Or do we? In the *Yan Tan Tethera* charm and all three of the *Mai-Dun* themes, this one-two-three-ness can be perceived, but at the end of the longer third segment, that horn call in two of the instances, the rising movement and developing logic fall back again in terms of musical pitch to leave us more or less where we started. This is even more true of *Legend*'s horn calls. Perhaps communication has failed.

It may fail for the moment, but what I think Ireland's three romantic orchestral works demonstrate is that while transcendent communication is not easy, is in fact increasingly difficult or contingent as 20th-century time proceeds, it is nevertheless possible. The three pieces were proposed as a trilogy by an early commentator. The observation is cited by Julian Herbage (1966), and may possibly have been his. Herbage considers them Ireland's 'three romantic orchestral works' by excluding the early *Tritons* and Poem in A minor, the works not for full orchestra (*A Downland Suite*, Concertino Pastorale), and the non-rural pieces (*Satyricon*, *A London Overture*, Epic March). The first, *The Forgotten Rite* of 1913, is a classic pre-war 'story of a panic', to quote the title of E M Forster's short story of 11 years earlier. A panpipe calls, and is almost inaudibly answered (Fig. 6.10 d, i). These calls are objects in a landscape, musical metonyms for Pan, but their development forms the shape and substance of the piece, in other words they become fused with the landscape itself as Pan's gigantic and libidinous presence transforms the entirety of experience, the process symbolised at the work's climax when Pan's motif heard as a horn sequence (Fig. 6.10 d, ii, bracketed motif *z*) leads to its ultimate fulfilment as trumpet calls prior to a classic dissolve. The capacity for transcendence of Ireland's little six-note motif would seem to prefigure by more than 70 years the similar function of a rather more famous five-note motif, that of *Close Encounters of the Third Kind*.

Mai-Dun and *Legend* are more complex and conflicted pieces, their soundscapes difficult to grasp fully. In the main section of *Mai-Dun*, repeatedly returning like a rondo theme, there is something strenuously physical, its score bristling with accents, dynamic markings and chunky agglomerations—the clash of weapons, the hefting of pickaxes, perhaps. Herbage points out that the composition predates by more than a decade Sir Mortimer Wheeler's claim (now discounted) that Maiden Castle had been a great battleground when taken by the Romans in AD 44, but his commentary on the music nevertheless seems wedded to a musical programme of repeated embattlement, appropriate enough to a rondo structure (not to mention the massive ring defenses of the monument itself): '*Mai-Dun* epitomises the strenuous life and struggles of a primitive community. The inhabitants of this early town-fortress must many times have had to repulse invasion before they finally succumbed to the Roman legions.' He was close to Ireland, so perhaps the composer did have this in mind. There is more, however. Three episodic events stand out. The first, using the themes of Fig. 6.10 c, ii and iii, seems contemplative or invocational but builds into an anguished cantilena of Mahlerian importunacy. The second shows that the main material of the work can also be playful, reminding us that it inhabits the *scherzo* genre. The third, near the end, is just a 'tranquillo' moment on a D flat major 6/4 chord, referring back to the initially empty landscape of *The Forgotten Rite*. After this, an effortful but decisive mood of triumph is attained in closing. The moment of emptiness is surely a crucial signifier of the very long view as a time-perspective on the work's subject.

The simplest interpretation of *Mai-Dun* with its chronological episodes would be to assert that if history is the meaning of a succession of events, then a piece of music, as a structured succession of sound events in which ideas recur or become transformed in time, can be an effective analogue to history. With a place as its title, *Mai-Dun* becomes not just the succession but the history, the structured, intellectual *meaning* of objects, including

a. Birtwistle, *Yan Tan Tethera*

b. Ireland, *Legend*

c. Ireland, *Mai-Dun*

i) Opening

ii) Second theme: transition

iii) Third theme: episode

d. Ireland, *The Forgotten Rite*

i) Opening

ii) Climax

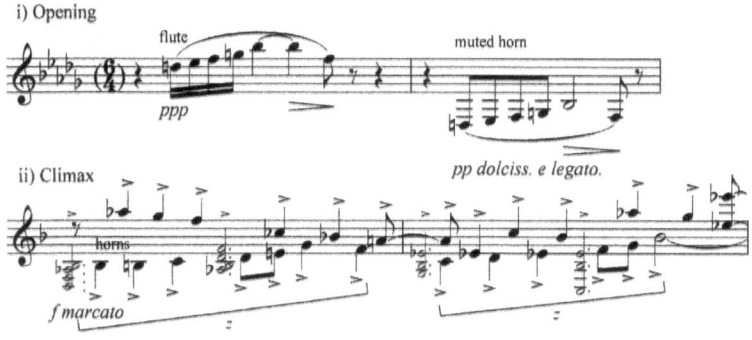

FIGURE 6.10: MUSICAL MOTIFS IN HARRISON BIRTWISTLE AND JOHN IRELAND

peoples and indeed the absence of peoples, passing across its landscape in time, through the ages. As its musical objects (themes) are related to each other, are transformed and contrasted, so the meaning of our relationship to the past, even the very distant past, is established. (See the bracketed *x* and *y* in Fig. 6.10 c, ii and iii, for an example of transformation of musical meaning.) At its most reductive, this might for example use a musical connection to indicate a genetic connection of peoples across the centuries.

Furthermore, as intensely subjective romanticism, Ireland's music has to be not merely a representation and interpretation of place and its actions across time, but a representation of how an individual can *experience* a place and the spirit of place across time. The implication is that if one responds, submits, to the music, one recreates the experience: will produces transcendence. One web author, in this same willing spirit, refers to the 4th-century AD Roman temple at Maiden Castle as having been built adjoining the site of 'an abandoned, but apparently remembered, circular Iron Age shrine' (Absolute Astronomy). Thus the rite is not forgotten at all, for the place somehow retains a memory. Music's job is to be able to recover it. Ireland himself, referring to the *Rite of Spring*, wrote: 'I always feel that the musical sounds Stravinsky makes in this work—the musical ideas themselves—seem to have the power of calling up something from the subconscious mind: some racial memory, perhaps, of things long hidden and belonging to a remote and forgotten past' (Herbage). *Legend* most fully expresses Ireland's belief, containing as it does a subjective presence in the orchestral site: a concertante piano, like a figure in the landscape. What strikes me most about *Legend* is that after the initial horn call, the invocation or invitation to experience, the long, stark piano solo for the first time in this trilogy suggests the difficulty of an uphill struggle, literally and spiritually. But on the plateau, once attained, the rewards are all the greater for submitting to the strenuous conditions of participation. Here, in a dance-like section, Ireland refers to an experience he had had on Harrow Hill: sitting there alone with a picnic, he was disturbed by a group of children dancing. When he looked again, they had vanished from the empty downland. He wrote to the occult novelist Arthur Machen about this, whose cryptic reply, on a postcard, was 'Oh, you've seen them too' (Herbage 1966).

Transcendent subjectivity
If we have strayed here from the strict topic of megaliths, we have at least remained with prehistoric sites and a modern subjectivity's relationship to them. Ireland's personal exploration in orchestral sound was of three such sites with its progression from a pure, unsullied landscape through an empty place with past events to a landscape with figures—or from a subjectivity, symbolised by Pan, completely at one with the landscape, through a vision of a past conflict, to the attempt of the alienated 20th-century individual to inhabit the vision. It seems to me that the ultimate point of these creations has to be, like the megaliths themselves, as a marker for something real that survives in a place over time. Or perhaps it survives in us in relation to place. If Rupert Sheldrake is right, memory does transcend genes and we shall somehow respond to a musical call from 5000 years ago when we hear it in a symphonic prelude in the concert hall. It is not, however, the only way of getting musical sense out of megaliths, and we can let John Pickard have the last word. I asked him in what sense, precisely, *Men of Stone* could be 'about' megaliths, and he answered with more eloquence than I, or even he, had anticipated:

> It's 'about' the megaliths at two levels . . . First there's the Romantic subjectivity . . . the sections are recollections of actual personal experiences of the sites . . . stone is generally very important to me: I come from a rugged area of Britain, dominated by millstone grit; my ancestors were stone masons; my dad and I used to go hunting for the tiny stone circles on the moors . . . so there's autobiography in there at some level. Second, there's the structural conceit of the circles within circles (times of day, seasons of the year), which I find enormously appealing as a way of structuring the music. There's even a tonal plan to the piece that runs through a circle . . . Now, the way they come together perhaps reflects the way those original megalith builders instinctively perceived their work: the building of those immense structures must have answered a deep emotional need in the first place, a need that is expressed through formality, calculation and social organisation. Which, for me at any rate, is as good a description as any of precisely what composing is all about. (Email communication, 30 October 2008.)

ACKNOWLEDGEMENTS
I wish to thank Peter Goodhugh for information about Thomas Lloyd Fowle.

CHAPTER 7

Stonehenge and its film music

Guido Heldt

Scoring the Stones
It seems safe to assume that its reception history has linked more and more different music, in a greater variety of ways, to Stonehenge than its active life as a place of ritual ever did. How films have scored Stonehenge is part of that history of reception and re-invention. But while the title of this text might be understood to imply that Stonehenge does indeed have its film music, a body of *topoi* forming a recognisable sonic image of the place or what it may mean, the evidence of my examples suggests a rather different picture, that of a bewildering range of solutions to the task of scoring the stones, latching onto different aspects of their reception history and their iconic status, or rather, their status as an icon for very different things.

This result has to come with a disclaimer. It is based on fiction films, not on documentaries, of which there are presumably far more, precisely because Stonehenge can be and has been used as an image for all sorts of things: for (the south of) England, for a (particularly northern European) ancient world, for that world's archaeological rediscovery and interpretation, for the very idea of the ancient, scientific and not-so-scientific mysteries, for the mysterious or the supernatural as such, for New Age religion, for particular kinds of rock music. The attempt to trace the musical presence of Stonehenge in cinema and TV documentaries across such themes and topics would not just be another paper but a research project in its own right.

Instead, this is a much more limited exploration of the musicalisation of Stonehenge (and a few other similar stone circles) in a number of fiction films and TV series across half a century. What emerges is a series of snapshots rather than a coherent and comprehensive picture; but in conjunction with other papers about Stonehenge reception history in this volume, it may contribute to a more meaningful panorama.

Tess of the d'Urbervilles
The most obvious place to look for film soundtracks engaging with Stonehenge is in adaptations of Thomas Hardy's *Tess of the d'Urbervilles*. On their flight from the police after Tess has murdered Alec d'Urberville, she and Angel Clare encounter Stonehenge not through sight, but through sound (see above, p. 20) and through touch:

> They had proceeded thus gropingly two or three miles further when on a sudden Clare became conscious of some vast erection close in his front, rising sheer from the grass. They had almost struck themselves against it . . . Lifting his hand and advancing a step or two, Clare felt the vertical surface of the structure. It seemed to be of solid stone, without joint or moulding. Carrying his fingers onward he found that what he had come in contact with was a colossal rectangular pillar; by stretching out his left hand he could feel a similar one adjoining . . . Tess drew her breath fearfully, and Angel, perplexed, said—
> 'What can it be?'
> Feeling sideways they encountered another tower-like pillar, square and uncompromising as the first; beyond it another and another. The place was all doors and pillars, some connected above by continuous architraves.
> 'A very Temple of the Winds,' he said.
> The next pillar was isolated; others composed a trilithon; others were prostrate, their flanks forming a causeway wide enough for a carriage; and it was soon obvious that they made up a forest of monoliths grouped upon the grassy expanse of the plain. The couple advanced further into this pavilion of the night till they stood in its midst.
> 'It is Stonehenge!' said Clare.
> 'The heathen temple, you mean?'
> 'Yes. Older than the centuries; older than the D'Urbervilles! Well, what shall we do, darling? We may find shelter further on.'
> But Tess, really tired by this time, flung herself upon an oblong slab that lay close at hand, and was sheltered from the wind by a pillar (501–2).

Hardy underlines the mystery of Stonehenge by making Tess's and Clare's encounter with it a mystery in itself. It is an uncanny place, and the two not so much reach it as they seem to be ambushed by it. Yet paradoxically, this simultaneously massive and open, wind-blown place becomes Tess's last place of shelter before the police—and by extension the entire tragic story of her life—catch up with her. Hardy's curious way of introducing Stonehenge is not only fascinating in the light of recent experiments concerning the soundscape of Stonehenge (see chapters 2 and 4 above), it is also intriguing in the way it counteracts, almost perversely, the fact that Stonehenge is iconic in the literal sense—it is an image, perhaps the strongest, most recognisable the history of this island has yet produced. Hardy denies us the image, and when we, with Angel, realise what that place is that they have stumbled upon, our mind supplies the image that is missing—that has to be missing in a novel that can evoke sensory experience only by verbal description and relies on our filling that description with (the memory of) experience.

It is an ingenious trick by which to deal with the limitations, the dryness of a verbal text, but it is one that sets a challenge for a screen adaptation. While film may not be quite the 'visual medium' that film scholarship sometimes likes to make it out to be, but has always had a strong audio component, it is still too strongly visual to trust that component enough to follow Hardy into his refusal of the image. And indeed, none of the three screen adaptations of *Tess of the d'Urbervilles* I watched for this text dares to do that: all three supply the image straight away. But all seem to try to find their own

FIGURE 7.1: *TESS OF THE D'URBERVILLES*, BBC MINI-SERIES (2008)

equivalent to Hardy's reduction of information. His Stonehenge encounter is (the description of) sound and touch, but not sight; the films give us sight, not much sound (with one exception), but no music, at least not for the moment when Tess and Angel and we set eyes on Stonehenge—a significant omission, as the moment has an impact that seems to call for music, and in classic Hollywood film language would have been musically underlined.

In the most recent version, the 2008 BBC mini-series, this refusal of underlining the moment with music is played out most clearly, because it leaves only a relatively brief (*ca*16″) music-less interval for the sighting of Stonehenge. The flight of Tess (Gemma Arterton) and Angel (Eddie Redmayne) from the mansion is scored with quick string ostinati. When they disappear behind a bush, the image cuts, and next we see Stonehenge in all its solitary glory on the plain—not really at night, as in Hardy, but in the falling evening light—before the two enter into the bottom right foreground of the picture (Fig. 7.1). For this, the music is silent, to give the image precedence; the wind is the only thing we hear. Only when Angel and Tess have got over the surprise of stumbling upon the stones and begin to walk towards their shelter does the music start again with slow-moving string chords and a soaring solo violin (that is, when the film leads us back to their story), in time-honoured fashion serving as a 'signifier of emotion' (Gorbman: 73).

Music fulfils a somewhat similar function in the equivalent scene in the 1998 London Weekend Television production of *Tess of the d'Urbervilles*. Here, too, the sighting of Stonehenge is stressed by the lack of music: Tess and Angel (Justine Waddell and Oliver Milburn) leave the mansion to the pastoral accompaniment of a slow siciliano melody (Fig. 7.2, the last phrase of the theme tune of the film first heard over the credits). But when the film cuts to an image of the moon in an ink-black sky, the music is replaced by the sighing of the wind, before the camera pans down to show us Stonehenge in close-up. The film takes a few seconds until we get a clearer view of the stone circle

FIGURE 7.2: ALAN LISK, *TESS OF THE D'URBERVILLES*, PASTORAL MELODY (AURAL TRANSCRIPTION)

FIGURE 7.3: *TESS OF THE D'URBERVILLES*, LONDON WEEKEND TELEVISION (1998), CLOSE-UP

FIGURE 7.4: *TESS OF THE D'URBERVILLES*, LONDON WEEKEND TELEVISION (1998)

FIGURE 7.5: ALAN LISK, *TESS OF THE D'URBERVILLES*, 'MEMORY' THEME (AURAL TRANSCRIPTION)

(Fig. 7.4), perhaps in an attempt to parallel Hardy's deliberate withholding of information; on the other hand, the stones are recognisable enough even in the initial close-up (Fig. 7.3).

More important is the disorienting effect of the spatial and temporal leap in the cut from pastoral landscape to night sky. Everything is aimed at this break in continuity, and consequently the music is not allowed to bridge the cut, as it does in other scenes. (The only remainder of standard musical continuity editing is the reverb of the last chord of the music echoing across the cut.) Instead, the lack of music underlines the extraterritorial space that Stonehenge becomes for a short while for Tess and Angel, and that is spelled out when Tess says that she loves the place: 'It's so solemn and lonely, with nothing but the sky above our heads. Seems as if there were no people in the world but we two.'

Here the gap is much wider, more than 2½ minutes long, and Tess and Angel are well into a dialogue about their relationship before the music starts again. It does so when Angel, just with a kiss, answers Tess's question about whether he believes they will meet again after they are dead, and a reduced variant of the siciliano theme underscores her desperate 'Angel, I fear that means no, and I wanted to see you again so much, so much' (Fig. 7.5). The music does not stress the shift from Stonehenge back to the lovers, as in the 2008 BBC series. Instead, that shift having happened earlier, the music re-enters with the realisation that the stone circle is not just a temporary shelter on their way, but the end of the journey—underlined with a non-tune that sounds like a memory of the film's pastoral siciliano theme, as if anticipating the fact that soon their story will be nothing but a memory.

The most unusual solution for the challenge of Hardy's Stonehenge encounter, at least on a sonic level, is to be found in Polanski's 1979 *Tess*. Visually, the film is close to the 2008 BBC TV version: we get the full image of Stonehenge, albeit later at night, in less light. The only visual concession to the mystery encounter Hardy sets up is the fact that the scene opens with a shot of the lovers' astonished faces before it shows us the source of their wonderment.

The more interesting part is the soundtrack, and the term is indeed relevant here. Again, music is absent, and does not come in during the Stonehenge scene. But in addition to the customary sound of the wind, the stone circle is filled with a strange soundscape of its own, mixed not very high, but high enough not to be missed: an indecipherable layer of moans and cries and rustles and squeaks. Going back to the novel, one could hear it as an attempt to realise the humming of the wind going through the stones, Hardy's 'booming tune, like the note of some gigantic one-stringed harp', an attempt to give the stone circle an acoustic presence in addition to its visual impact. But the soundscape the film creates does not come close to Hardy's description, nor to a realistic sonic image of the sound of the wind in a circle of close-standing stones. It seems less natural, rather like something emanating from the stones; one could imagine it as a memory of Stonehenge's long history, a layering voices sedimented over the millennia.

The indecipherability of the sounds is a double one, with regard both to their actual, empirical sources and to the putative source in the story world of the film. Are we

FIGURE 7.6: DAVID MILNER ED, *THE HIGHWAYS AND BYWAYS OF BRITAIN*, COVER

listening to more or less distorted animal noises? To electronic sounds? To a mixture of both? And do these represent a natural phenomenon, or the magic of the place? Hardy's Stonehenge acquires its booming tune from the interplay of nature and culture, and this interplay is a fitting image for the in-between space that Stonehenge is in the story: a space open to the elements, yet the last shelter Tess finds on her flight; a monument of ancient history, yet also a place one of whose stones the exhausted Tess uses as a bed; a place of sacrifice foreshadowing the sacrifice that will be made of Tess to preserve the current fantasy of social and moral propriety, but also the place of her last happy moments with Angel. The in-between soundscape the film finds for Stonehenge, while not in any way close to Hardy's description or to acoustic realism, is nevertheless an effective parallel to the status of the place in the story, and a way of avoiding the musical clichés to which both TV productions cannot refrain from falling prey.

Highways and byways
The music connected to Stonehenge in a film necessarily is a function of the stones' function in it, usually taken from a limited repository of uses and clichés. The dust jacket of a recent book (Fig. 7.6) shows one of them. Here it is set in a prototypically green and pleasant southern English landscape, one that does not look very much like Salisbury Plain but largely conforms to the idealised landscape of late 19th- and early 20th-century

FIGURE 7.7: *SHANGHAI KNIGHTS* (2003)

FIGURE 7.8: RANDY EDELMAN, *SHANGHAI KNIGHTS*, PASTORAL MELODY (AURAL TRANSCRIPTION)

pastoral visions of England's south (see Howkins). It is a touristy image, and the culture of countryside excursions developing in the 1920s and 1930s both relied upon and furthered the popularity of this image (Howkins: 83). Indeed, the book is a re-issue of excerpts from Macmillan's *Highways and Byways* series of books, originally published in the first half of the 20th century as guides to places of beauty and interest around Britain. Nature, culture and history seamlessly flow into one in the illustration, and the road with the two white lane-dividing stripes in the foreground promises comfortable accessibility of these places. To achieve that seamless integration, Stonehenge has to be transported into the 'wrong' landscape of rolling hills, an inaccuracy one could critically see as a shameless fabrication, or more charitably as a tongue-in-cheek way of admitting that what we are presented with is a fantasy.

Remarkably, filmic versions of a peacefully pastoral Stonehenge seem to be rare. The only proper example I could find, though, goes much further than the *Highways and Byways* illustration in ostentatious cheekiness. The scene in question occurs in *Shanghai Knights* (2003), a sequel to *Shanghai Noon* (2000). *Shanghai Knights* sees Chinese Chon Wang (Jackie Chan) and American Roy O'Bannon (Owen Wilson) travel to England to recapture the stolen Chinese Imperial Seal, and they do not just succeed in their endeavour but uncover a conspiracy to murder the British royal family into the bargain. An adventure comedy made by Hollywood featuring heroes from far-flung corners of the globe travelling to Victorian England cannot but overdo the England clichés, and the scene involving a fantasy version of Stonehenge is only one of a series of these.

The impact of pastoral Stonehenge is reinforced by the fact that the scene follows a daydream in which O'Bannon imagines himself at the receiving end of the sexual predatoriness of Fann Wong (Chon Lin), only to wake up to the fact that the tongue licking his face belongs to a sheep that has clambered into the broken-down car in which he is sitting. The sheep is an integral part of a landscape that bears a striking resemblance to the one from the *Highways and Byways of Britain* dust jacket (Fig. 7.7). The pseudo-Stonehenge is sufficiently close to be recognisable as what it is supposed to be, but sufficiently inauthentic to remain an obvious fantasy; that the scene is a parody both of clichéd images of a quaint old England and of ignorant foreigners staring at it is rammed home when Chon asks, 'Who would leave a pile of stones in the middle of the field?' and Roy answers, nonplussed, 'I don't know, Chon—these people are nuts.' Musically, the gulf between tourist-guide England and Roy's erotic fantasy is realised by the switch from The Zombies' 'Time of the season' (*Odessey and Oracle*), as the music for his dream, to a sweetly pastoral flute melody above guitar accompaniment that starts with the shot shown above and develops when the two wander through the English landscape, eventually also involving piano, harmonica, a harp glissando and strings (Fig. 7.8).

Another way of giving Stonehenge a positive slant is to picture it—visually and musically—as a place of healing (which also seems to be one of the current archaeological theories). An example occurs in *The Tomb of Ligeia* (1964), rather loosely based on Edgar Allan Poe's story 'Ligeia'. The story does not mention Stonehenge at all, and in the film it occurs only once, rather ornamentally, when Verden Fell (Vincent Price) is wandering about the landscape with his new wife Rowena (Elizabeth Sheperd) shortly after the marriage, and they are talking about their new life together and their hopes for the future, soon to be dashed by Fell's increasingly vivid conviction that his first wife Ligeia (also Elizabeth Sheperd), who had promised never to die, is returning and possessing Rowena. During their wanderings, the couple also pass by Stonehenge (which here seems to be close to the sea, as just before we see them walk on a beach), and Verden tells Rowena that 'in Celtic religion, Stonehenge was a temple to the god of healing. It was built more than 3,000 years ago, and do you know why it remains today, Rowena? Because it was built with a sense of purpose. Stone by stone, like the pyramids in Egypt, or like the Aztec towers in Mexico' (Fig. 7.9).

We assume that he himself is hoping for healing, for a new purpose in his life, away from the baleful shadow of Ligeia, and the music supports his hopes with yet another sweetly meandering flute melody over a gurgling harp accompaniment. But Rowena unwittingly steers their dialogue back to the fateful future when she asks, following Verden's comparison of Stonehenge with the pyramids and Aztec towers, 'Like our abbey?' and he answers, 'Yes, like our abbey, my dear.' That Verden lives in a dilapidated abbey, and that it is here that he and Rowena are increasingly haunted by Ligeia or by his memories of her, may make the inclusion of Stonehenge significant, because one could understand it to be meant to set up a contrast between quiet, timeless, pagan Stonehenge as a place of healing and the morbid, romantically charged atmosphere of the ruined (Christian) abbey, though the film does not belabour the point. In Poe's story Verden Fell makes it clear that the choice of the abbey as his dwelling place was a direct result of his despair at having lost Ligeia: 'After a few months, therefore, of weary and aimless wandering, I purchased and put in some repair, an abbey, which I shall not name, in one of the wildest and least frequented portions of fair England. The gloomy and dreary grandeur of the building, the almost savage aspect of the domain, the many melancholy and time-honoured memories connected with both, had much in unison with the feelings of utter abandonment which had driven me into that remote and unsocial region of the country' (357).

A rather different way of using this idea of ancient stone circles as places of healing is exploited in *Quatermass* (1979). The TV mini-series, based on three 1950s predecessors and a cinema film from 1967, all with the same main character of Professor Bernard Quatermass (John Mills in the 1979 incarnation), provides a strangely belated reaction to hippie culture. We see the 'Planet People' gather in large throngs in different places all around the world, several stone circles among them, in the hope that they will be transported to a different planet, their imagined real home, away from degenerate earth.

The film uses an actual photograph of Stonehenge for its credits, but then builds its own invented stone circle, Ringstone Round, that even comes with its own nursery rhyme:

> Huffity, puffity, Ringstone Round,
> If you lose your hat it will never be found,
> So pull your britches right up to your chin,
> And fasten your cloak with a bright new pin,
> And when you are ready, then we can begin,
> Huffity, puffity, puff!

FIGURE 7.9: *THE TOMB OF LIGEIA* (1964)

FIGURE 7.10: NIC ROWLEY AND MARC WILKINSON, *QUATERMASS*, CHORALE (AURAL TRANSCRIPTION)

FIGURE 7.11: *QUATERMASS* (1979)

(It is generally assumed that the nursery rhyme was written by script writer Nigel Kneale for the series and its fictitious stone circle, but the matter does not seem to be quite so clear-cut, and a real-life origin for the place name and the rhyme have been suggested. See Mama Lisa's World: England). For the Planet People, the stone circle is a holy place, and while they themselves just shout and make noise on little bells and other percussive instruments when they are finally approaching Ringstone Round, the nondiegetic soundtrack gives a musical voice to their hopes with a slow-moving chorale (Fig. 7.10). The religious transport and literal transportation to another planet that they hope for at the ancient holy site seems to take place when a blinding light appears in which everyone disappears—unfortunately not to the dreamt-of home planet, but to that of aliens who use the poor deluded souls as a food supply. Here the music changes tack and replaces the chant with electronic textures ending in hissing and tinkling sounds for the images of the ash-strewn stone circle after the light has vanished again. The music for the appearance of the light and its aftermath (Fig. 7.11) hovers between diegetic and nondiegetic localisation. It could be understood as an attempt to suggest the sounds of the extraordinary event; it could also be understood as a way of musically evoking or underlining its shock and intensity. But whichever way we hear it, what is important is the switch in perspective: the chant voices the Planet People's own subjective perception of the situation, but with the light the music abruptly returns to harsh reality and (at least eventually, when we learn what the light signified) the stone circle emerges as a place not of healing but of death.

FIGURE 7.12: *NIGHT OF THE DEMON* (1957), OPENING IMAGE

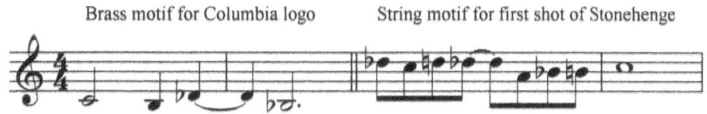

FIGURE 7.13: CLIFTON PARKER, *NIGHT OF THE DEMON*, CHROMATIC MOTIF (AURAL TRANSCRIPTION)

In *Quatermass*, this flip to the dark side is an indictment of the starry-eyed naivety of New Age religiosity, but the idea of Stonehenge as an ominous place, even a place of evil, has its own tradition—in art history with images like William Overend Geller's successful mezzotint *The Druid's Sacrifice* (1832), but also in film history, prominently at the beginning of *Night of the Demon* (1957). It is an odd opening: the film is based on a story by M R James, 'Casting the runes' (1911), and as is usually the case in James's stories, it builds up its horror slowly and carefully, beginning with a few letters outlining the refusal of an unnamed association to accept a paper on alchemy for one of its sessions and an almost perfunctory discussion about the matter between the secretary of the association and his wife. The film, on the other hand, begins with an image of Stonehenge below a leaden sky (Fig. 7.12) and a portentous voiceover informing us that 'it has been written since the beginning of time, and even unto these ancient stones, that evil, supernatural creatures exist in a world of darkness. And it is also said, man using the magic power of the ancient runic symbols can call forth these powers of darkness, the demons of hell. Through the ages, men have feared and worshipped these creatures; the practice of witchcraft, the cults of evil, have endured, and exist to this day.'

If the opening of the film goes for all-out Lovecraftian cosmic horror, it does an injustice not just to M R James but also to the rest of the film, which is—apart from an ill-judged early appearance of the demon of hell—far more subtle, and more in keeping with the tone of James's story. But the music follows suit and ratchets up the horror even more. Already the Columbia company credits are accompanied by a heavy, chromatic four-note brass motif. For the Stonehenge shot underlying the voiceover, the music, after a gong stroke, uses another chromatic motif, now in the high strings (Fig. 7.13). Subtlety does not seem to be among the things Stonehenge inspires, certainly not in film, whether pastoral or ominous; *Night of the Demon* needs quite a while after this opening to become the generally fine little horror film it is, not up to the standards of the films Tourneur directed for producer Val Lewton at RKO in the 1940s, such as *Cat People* or *I Walked with a Zombie*, but nowhere near as crass as its credit sequence.

FIGURE 7.14: *MUPPETS FROM SPACE* (1999)

FIGURE 7.15: *EVIL ALIENS* (2005)

Letting the air out and the music in

Be it as a place of the pastorally picturesque, of healing, or of horror and evil, what is never in doubt is the importance of Stonehenge, its significance as an icon, whatever it may be an icon for. Another strand in the filmic reception of Stonehenge has attempted to let some air out of the frequent pomposity of this unspecific significance and to poke fun at the status of the stones. That is the case in two recent screen appearances of Stonehenge.

In *Muppets from Space* (1999), Stonehenge, quite in line with the Planet People's fantasy in *Quatermass*, becomes a means of communication with the cosmos, but here the cosmos informs us and the muppets merely that Gonzo is in fact an alien and is about to be visited by his family. Stonehenge (Fig. 7.14) is one of a series of sites of messages from space—the pyramids, the Hollywood sign (including an extraterrestrial spelling mistake), Gonzo's breakfast cereal. The comedy *Evil Aliens* (2005) invents its own stone circle in a remote corner of Wales, which turns out to be a docking port for alien spaceships; but briefly Stonehenge itself features. The film is a mishmash of alien horror and media satire, the latter being played out around cable TV programme *Weird Worlde*, which caters to those interested in UFO sightings, alien abductions, paranormal phenomena etc, and which is headed by deeply cleavaged and deeply cynical presenter Michelle Fox, seen in Fig. 7.15 with her sleazy producer, Stonehenge on a poster in the background, and a badly photoshopped flying saucer hovering above it.

Neither of these films does musically interesting things with its Stonehenge. That, of course, is also true of other films, for example *Fiddlers Three* (1944), where Stonehenge is the conduit that allows the three heroes to time-and-space travel to ancient Rome and back, but without culturally or structurally noteworthy results in the film score. An example of pop culture that does offer musical interest also works to demolish the iconic status of the stones, but in a way that paradoxically reinforces it. In the Beatles' *Help!* (1965), the four need to be protected from an eastern cult attempting to sacrifice the bearer of its ceremonial ring, which has found its way onto the finger of Ringo. Eventually, they have to relocate a recording session onto Salisbury Plain, where they play surrounded by British tanks while the cult members gather around and wage battle against Britain. The choice of location is justified by the fact that parts of the plain are indeed used for military purposes, while it also provides the opportunity for a cultural sea-change. Stonehenge features in the scene only once, almost perfunctorily, far in the background (Fig. 7.16). But we also get a shot of the Beatles from above, in the centre of a circle of tanks that looks suspiciously like a stone circle (Fig. 7.17). The juxtaposition of images unmistakably installs the Beatles as the new cultural icon and national treasure, a treasure that even comes with its very own music and saves the film the attempts to find a musical idiom for the monument.

Of the films mentioned in this brief survey, *Help!* is by no means a particularly late one; only eight years separate it from *Night of the Demon*, and only one from *The Tomb of Ligeia*, though in the 1960s a short time could mean a considerable cultural gulf. But as more recent film(-musical) appearances of Stonehenge from *Quatermass* to *Tess of the d'Urbervilles* and *Shanghai Knights* show, dethroning or deconstructing Stonehenge is hard to do, and its myth and its many cultural uses live stubbornly on.

FIGURE 7.16: *HELP!* (1965), STONEHENGE IN THE DISTANCE

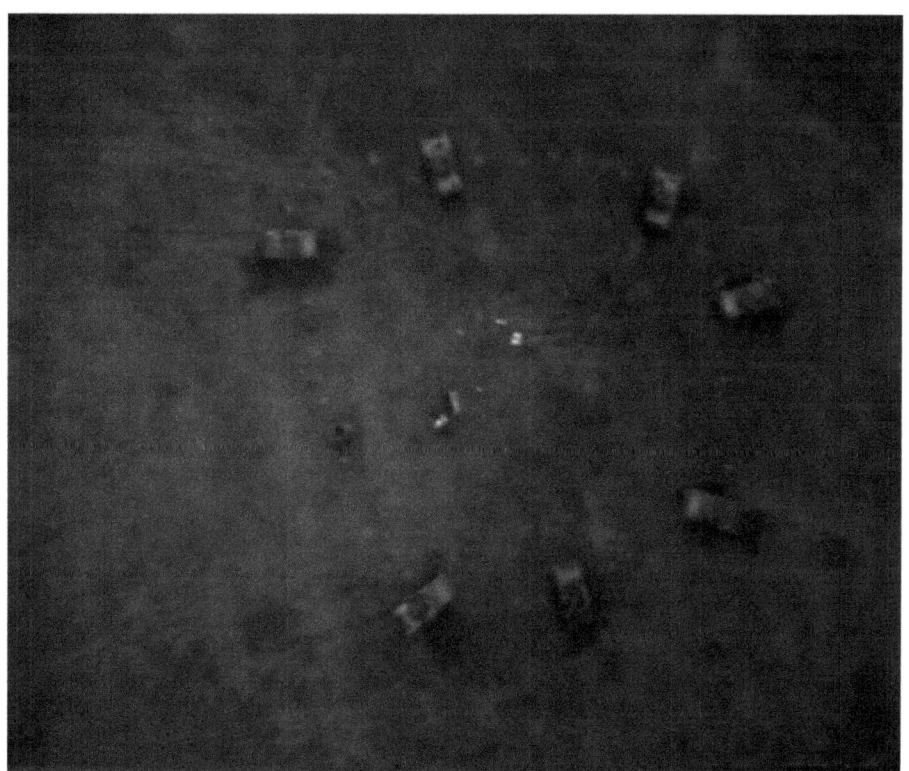

FIGURE 7.17: *HELP!* (1965), AERIAL SHOT

CHAPTER 8

Stonehenge in rock

Timothy Darvill

Sticking out of the hallowed turf of Salisbury Plain like the wickets of some long-abandoned cricket match are the stone stumps and precariously balanced bail-like lintels of the world's most famous prehistoric monument: Stonehenge (Fig. 8.1). Unlike many great archaeological sites, Stonehenge did not have to be discovered in the way that Howard Carter found the tomb of Tutankhamun, Chinese archaeologists opened the serried ranks of the Terracotta Army at Xi'an, or Giuseppe Fiorelli painstakingly uncovered Pompeii. Like the Pyramids of Giza, Stonehenge has been a powerful feature in the landscape since the first stones were raised around 2300 BC. We find it mentioned in the earliest written accounts of southern Britain by geographers and historians. The Greek historian Hectataeus of Abdera, for example, writing in about 330 BC, speaks of a temple in the land of the Hyperboreans. If we fast-forward to medieval times, Henry of Huntingdon in about AD 1130 describes the site as 'having been erected after the manner of doorways', and from the same period Geoffrey of Monmouth explains Stonehenge as a memorial to noble warriors slain in battle (Darvill 2006: 32–6). Stonehenge is by far the best documented archaeological site in Britain, with a bibliography listing thousands of texts and illustrations. Yet it still provides mystery and intrigue. Small wonder, then, that for centuries Stonehenge and similar kinds of ancient monument have been a stimulus, a stone muse, to the imaginations of writers, artists, and musicians great and small.

This chapter focuses on one dimension of the way Stonehenge has impacted on the arts and popular culture: its role in the world of rock music through the later 20th century and a little beyond. Naturally, such a relationship has a wider context (Darvill 2004a), but, as we shall see, the image of the upright stones and their lintels crosses cultural and musical boundaries in much the same way that rock music in its broadest sense does too. Stonehenge's instantly recognisable form allied to easily remembered beats is a powerful cocktail that can reach deep into people's souls and, for musicians and promoters alike, deep into the pockets of the paying public. At the front end of the relationship between music and the monument are amusing wordplays that turn a few heads and raise a smile. In summer 2008, for example, South West Trains used an advertising campaign with posters showing a stripped-down image of Stonehenge with flowers around the pillars and the caption 'See the Stones'. Hinting at an ambiguity between Britain's most visited ancient monument and hearing Britain's most famous ancient rock band, the Rolling Stones, fresh from their appearance at the Isle of Wight Festival, the company obviously hoped to whisk customers off to both these places and others beside. And topping them all, English Heritage chose the soundbite 'Stonehenge Rocks!' as the theme for their 2008 promotion and merchandising campaign, with those immortal words emblazoned on knick-knacks from T-shirts to chocolate bars, and in deference to the spirit of the age they even resurrected sew-on patch-badges for your Levis (Fig. 8.2, colour).

Four aspects of Stonehenge in the world of rock and roll are considered here, together highlighting the reach and power of the monument for the cultural reception of music and all its trappings. After a recognition of Stonehenge as the inspiration for bands and their music, the spotlight moves to the use of the Stonehenge image in artwork associated with the packaging and delivery of music. From there it is short step to Stonehenge as a backdrop for the performance of rock music, and finally to the use of Stonehenge itself as a venue.

Band names, songs and album titles

A search of the internet will quickly reveal a clutch of bands using Stonehenge in their name, and dozens of album titles and song tiles that incorporate or reference the name of the site in one way or another. Among the bands that style themselves after the site is the Birmingham-based British reggae group simply called Stonehenge, featuring Dennis Bovell and Errol Pottinger. In 1970 they changed their name to something a bit more in keeping with their music—Matumbi—and had some success with a style of reggae dubbed 'lover's rock' before disbanding in 1982. Across the Atlantic, a band calling themselves the Druids of Stonehenge started in New York before moving to Los Angeles in the mid-1960s, playing a variety of rhythm and blues overtly modelled on the success of the Rolling Stones, and perhaps picking up on the rock–stones–Stonehenge connection in selecting their name. Their most recent eponymous vinyl EP (2008) picks the best of their back catalogue and well illustrates their energetic brand of driving rock. Slightly more contemporary in feel is the work of a collective known as Stonehenge based on collaborations with Canadian singer Wendy Grondzil. Their release *Remember me?* (2008) is promoted on the cover-notes as 'a torrent of emotive music spanning multiple streams of consciousness' alongside a picture of what looks like prehistoric rock art.

Albums called *Stonehenge* and variations thereon abound. One of the earliest and most powerful is *Stonedhenge* by the rhythm and blues-influenced band Ten Years After fronted by Alvin Lee (1968). To emphasise the connection, the folding cover shows a painting of the surviving northeast sector of the sarsen circle at Stonehenge with a stylised midsummer rising sun superimposed on a landscape covered in Beardsley-esque beasts and people (Fig. 8.3, colour). The name of

FIGURE 8.1: STONEHENGE, WILTSHIRE. VIEW OF THE CENTRAL STONE SETTINGS LOOKING SOUTHWEST
(AUTHOR'S PHOTO)

the album was later used by Christopher Chippindale (1986) as the title for his review of events surrounding the 1985 summer solstice celebrations at Stonehenge to which we shall return later.

Title and image often go together. Richie Havens used a photograph of the southwestern trilithon (stones 57, 58 and 159) after its re-erection in 1958 with the sun shining through the jagged gap between the uprights on the front of *Stonehenge* released in 1970. Likewise, Hawkwind, a band with many connections to Stonehenge (see below), released *Stonehenge: This Is Hawkwind Do not Panic* in 1984 with drawings of Stonehenge in wide view across the cover. Hiding amongst some fairly lengthy tracks are 'Stonehenge decoded' after the book of the same title by Gerald Hawkins, and 'Circles' based on the configuration of stone settings at the site.

More detached is the album *Stonehenge* by Nijssen and Vanloan (2007) with a rather plain cover but music that claims to 'bring the spell of Stonehenge to life', according to promotional notes on the CD Baby website. Confusing early medieval script with the prehistoric past seems to be the ticket for New Age band Runestone, who released an album called *Stonehenge* in 1992.

Songs about Stonehenge, and music inspired by the site, regularly turn up in various degrees of isolation from the physical remains of the site itself. A selection illustrates the range. Black Sabbath included a song entitled 'Stonehenge' on their album *Born Again* (1983), issued to accompany a tour of Europe and North America in 1983-4 which, as we shall see, had a Stonehenge theme. The album was panned by the critics, but with a loyal fan-base reached no 4 in the UK charts. A single of the track 'Stonehenge' was released but, perhaps not unexpectedly, failed to chart. In a quite different style, Kellianna has the song 'Stonehenge' on her album *Lady Moon* (2004), while in the same genre Gregory claims a visit to Stonehenge as an influence for the composition 'Sunrise over Stonehenge' on *Circle of Time* (2008). The Dorset-based folk band the Dolmen include a song called 'Stonehenge' on their self-promoted album *Ah Ry Ah* (2008), and have been known to play at Stonehenge itself during neo-Druid events and ceremonies. Different again, the highly political Norwich-based band favouring direct action known as the Disrupters included a track called 'Stonehenge' on their compilation album *Gas the Punx* (2005), written in response to riots connected with access to the site in the mid-1980s (see below). And in a quite different style again, reflecting an intimate knowledge of the source of some of the stones used in the construction of the monument, the song 'Preseli bluestone: the Stonehenge crystal' by Amie Ridley can be found on her album *Crystology* (2008).

Artwork and image
Every bit as important as the names of bands and the music they create is the packaging in which the product is delivered into the listener's hands. Traditional album

covers of the 1960s, 70s, and 80s containing 12-inch vinyl LPs were exactly 12⅜″ square and are widely recognised as being among the most distinctive, most widely reproduced, and most viewed works of art of the later 20th century. Literally, millions of people have seen the cover artwork of the most popular releases even if they don't own a copy. As Storm Thorgerson and Aubrey Powell point out, 'a cover design is the icon that identifies—and is invariably identified with—the music it represents' (9). With the demise of vinyl LPs and the rise of CDs, something of the power of the album cover has been lost, but the smaller format retains more or less the same shape and even at its diminished scale still provides an interesting canvas for the designer, artist, and photographer to work with.

Simultaneously inspiring and enigmatic, well in keeping with the progressive rock tradition that it represents, was the cover of the Yes album *Tales from Topographic Oceans* (1973) and especially the inner face of its folding double spread (Fig. 8.4, colour). For this image, landscape artist Roger Dean created a panorama of a mythical prehistoric world, acknowledging reference to John Michell's book *The View over Atlantis* (Michell 1969). In the background is the sun rising over a Mayan pyramid from Chichen Itza (Mexico) and geoglyphs from the Plain of Nazca (Peru), while in the foreground are rocks based on images of stones at Avebury and Stonehenge (Wiltshire), Brimham Rocks (North Yorkshire), Last Rocks at Land's End (Cornwall), and Logan Rock near Treen (Cornwall).

Juxtaposing ancient sites from around the world is not unusual on album artwork and posters. The packaging for CD1 of *Year 2000: Codename Hawkwind. Volume One* by Hawkwind (1999) has a general view of Stonehenge just below centre with a group of four Easter Island *moai* statues bottom left, and a sky dominated by the sun and star-like images of the band-members' faces. Similarly, Stonehenge was depicted alongside rock formations in the Nevada desert, cacti, and topiary on the poster promoting Jools Holland's winter 2006 tour supported by his Rhythm and Blues Orchestra and guest vocalists Lulu, Ruby Turner, and Louise Marshall. The poster was also used as the cover for the 2006 album *Moving out to the Country*.

Hard and soft images of Stonehenge crop up in all corners of the rock music world and its associated genres. Heavy metal rockers Aerosmith have a strangely distorted view of Stonehenge on the front of their album *Rock in a Hard Place* (1982) but more conventional pictures elsewhere on the cover. A compilation of nearly 80 hard-riffing rock anthems released as *Rock of Ages* (2002) shows three spookily-lit trilithons on its cover, while the circles appear in long shot on *New Dawn* by Del Bromham and his band Stray (1998). More unusual, but in a way completely understandable, was the use of an image of Stonehenge on the cover of the Japanese version of the album *A* (Andy Scott's Sweet, 1995) released when Andy Scott reformed a version of 1970s glam-rock band Sweet to tour the far east.

At the folky end of the rock-music spectrum, a compilation of Irish flute and piano music entitled *Misty Celtic Morning* (2001) has Stonehenge on the cover, perhaps reflecting a medieval tradition that the stones for for the monument originated in Ireland, while a plan of Stonehenge is used to interesting effect on early copies of flautist/composer Tim Wheater's album *Sound Medicine Man* (2004). James McCarty uses a wide view of the site on the cover of *Out of the Dark* (1994), while Gary Sill and Bob Day have an exploding fire-burst almost totally eclipsing the monument on *Stonehenge is Burning* (2007), a strange way to celebrate what the promotional material bills as 'the very best of environmental and landscape jazz'. Overall, the variety of ways in which Stonehenge is treated by artists and designers is awesome and sometimes bizarre. Perhaps most extreme is an image from the far outer fringes of the material considered here: the American stand-up comedian Basil White who goes well beyond good taste with an album entitled *Peeing on Stonehenge* (2005), its cover image showing him doing just that.

But it is not only the original Wiltshire Stonehenge that appears. The cover and booklet notes of Steely Dan's *Remastered: The Best of Steely Dan* (1993) features views of the Carhenge sculpture by Jim Reinders, placed in a field at Alliance, Nebraska, USA, in 1987 (see Chippindale 1987: 23).

In all these works, and others that these few examples represent, it is the popularly perceived mysterious nature of Stonehenge matched by its instantly recognisable form that makes the image work. Much of the associated music from recent times is grounded in earlier traditions, survivals from a past, perhaps as hard to remember in the case of psychedelic rock as the meanings of Stonehenge itself are difficult to unravel. Subcultures that relate most closely to such images are very often interested in (even consumed by) science fiction, New Age philosophies, alternative existences, flying saucers, ley lines, paganism, and supernatural forces. Such things are part of the package, a point rather neatly summed up in the artwork of a poster advertising the two-day Sonic Rock festival held near Skegness, Lincolnshire, in September 2005 (Fig. 8.5, colour). Here there is more than a hint that Stonehenge is important as a physical place that provides the backdrop to other-worldly experiences and unusual adventures.

Stonehenge the backdrop
Being at Stonehenge instantly links rock bands and their brand of music with a heritage that even the sharpest managers could not manufacture. On their visit to Britain the mid 1970s, the Los Angeles-based jazz fusion band Spirit used Stonehenge as the backdrop for a series of widely published promotional pictures (Buckley *et al*: 937). New Ager Kellianna has a rather stylised close-up of herself standing in the entrance to the sarsen circle on *I Walk with the Goddess* (2007). The David Bacha Band look frozen standing in front of a trilithon on the cover of *No Sleep after Stonehenge* (2008). And to cap them all, a compilation of folk-songs under the title *Stonehenge: Mystic Cycle* (2000) has three oversized lute-like stringed instruments nonchalantly propped on the stones through the trickery possible within some computer-aided photo-montaging package.

FIGURE 8.2: STONEHENGE ROCKS! SEW-ON CLOTHING PATCH FROM THE RANGE OF ENGLISH HERITAGE MERCHANDISE INTRODUCED IN 2008 (AUTHOR'S PHOTO)

FIGURE 8.3: TEN YEARS AFTER, *STONEDHENGE* (1968)

FIGURE 8.4: YES, *TALES FROM TOPOGRAPHIC OCEANS* (1973). INNER COVER ARTWORK BY ROGER DEAN.

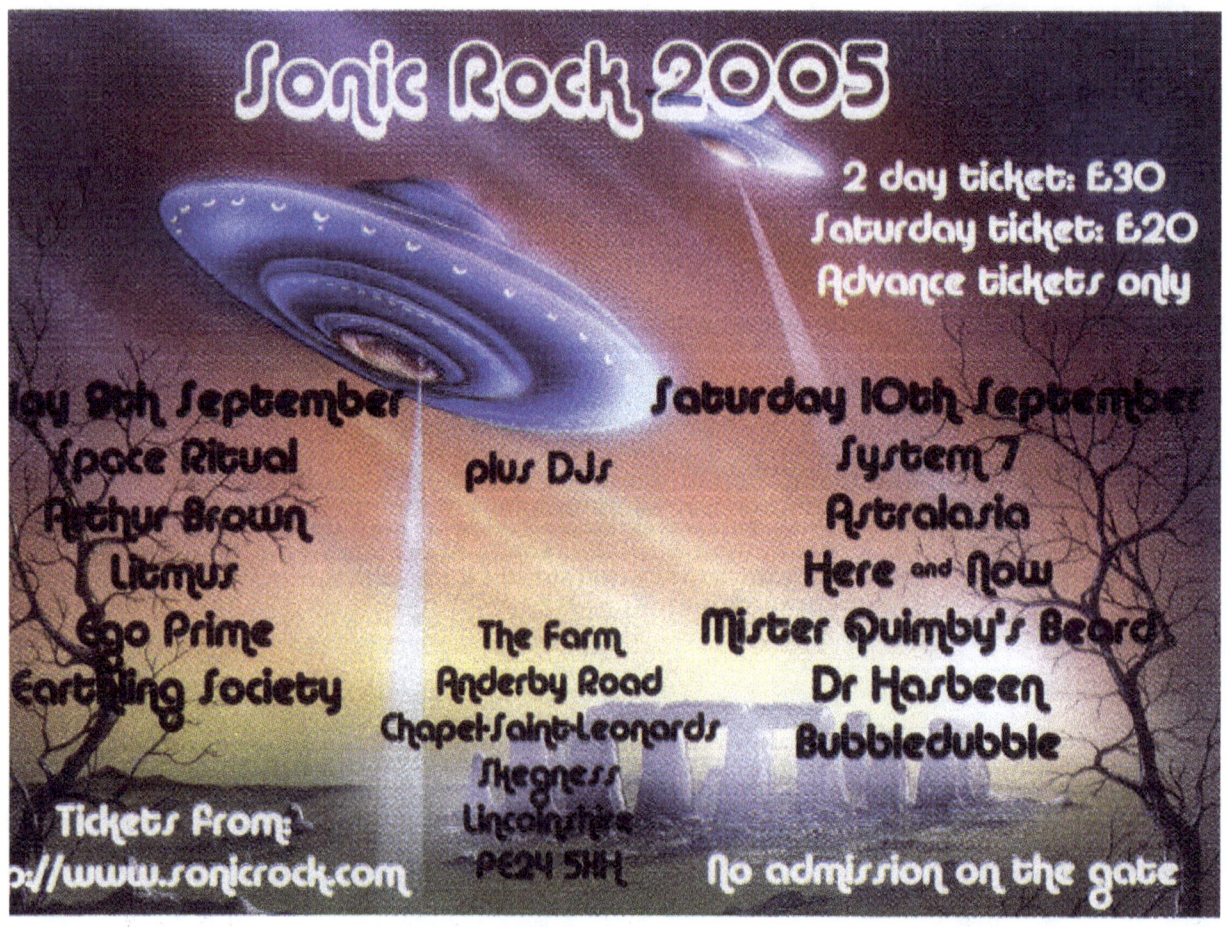

FIGURE 8.5: MAGAZINE ADVERTISEMENT FOR THE SONIC ROCK 2005 FESTIVAL FEATURING STONEHENGE AND A FLYING SAUCER. REPRODUCED FROM PROMOTIONAL ARTWORK

All these images are presumably a way of engendering familiarity on the part of the public, if not with the band themselves then at least as a marker for the kind of music listeners might expect and the sort of people who should buy the product. In rare cases, however, it is the other way round, as some of the biggest acts in the rock pantheon image-check the site. The Beatles are shown playing on Stonehenge Down, surrounded by tanks and military protection but with the stones clearly visible in the background, in the film *Help!* (Jones; see also chapter 7 above). Not to be outdone, Mick Jagger and Keith Richards of the Rolling Stones dutifully lazed around at Stonehenge in 1967 while Michael Cooper photographed them for the cover image of their psychedelic rock album *Their Satanic Majesties Request*, released that same year but without any trace of Stonehenge.

Of course, Stonehenge can only be where it is, so if you can't go to Stonehenge itself, why not re-create it? Led Zeppelin are one band that did just that, using a wooden Stonehenge stage-set for a concert at Oakland Coliseum, California, on 23 July 1977, and other dates on their North American tour. A few years later Black Sabbath used a similar gimmick on their 'Born Again' tour of Europe and North American between August 1983 and March 1984, but the main components were so large that at many venues it was impossible to get them onto the stage. Their interest in Stonehenge was registered, however, and some years later when a tribute album for the band was recorded it was titled *Hail to the Stonehenge Gods*.

More adventurous, and perhaps somehow referring back to the Zeppelin and Sabbath extravaganzas, was the use of Stonehenge in the satirical 'rockumentary' film *This is Spinal Tap* (1984—see Jones). Directed by Rob Reiner, this film still has a wide following and considerable popular acclaim. It portrays the fictitious heavy metal rockers Spinal Tap as a British band making a not very successful American comeback tour. Along the way they hit on the idea of using a stage set based on Stonehenge as a way of raising their popularity when interest in them starts to flag. A potentially great idea is thwarted when their stage designer creates a replica trilithon only a couple of feet high. Lowered onto the stage at the climax of the show, with the song 'Stonehenge' blasting from the speakers, it is dwarfed by the band members themselves in their platform shoes and big hair. The original sequence was filmed at the Shank Hall in Milwaukee, where the image of the prop is still used as the venue's corporate logo, and the cover of some versions of the album *This is Spinal Tap* (1984) carried an image of the model. In tribute to Spinal Tap, the Mancunian dance music collective the Happy Mondays used a similar diminutive trilithon on stage during an appearance at the Glastonbury Festival in June 1990, a stunt that journalists dubbed 'spinal crack' because of the band's reported heavy drug usage at the time (Gordon: 14).

Stonehenge the venue
No stage prop or image, however good or amusing, can beat the atmosphere and environment of Stonehenge itself as a dramatic and imposing place for a show. Whether or not the original design and structure of the site itself incorporated elements that enhanced the soundscape, there is no doubt that music played a part in annual gatherings to celebrate the summer solstice through the 20th century (Stout; Worthington 2002, 2004).

Rumours abound that Syd Barrett of Pink Floyd and Keith Moon of the Who together performed a song called 'Sun Ra' at Stonehenge on the 1969 summer solstice, but there is no substantive verification of the event. Perhaps it is kind of rock legend developed to provide a prehistory for what happened five years later when music fan Wally Hope and others promoted the idea of a 'Free Stoned Henge Festival' to coincide with the solstice in June 1974. Flyers by Phil Russell and airplay on Radio Caroline attracted a crowd of about 500, the name of festival perhaps recalling the title of the album *Stonedhenge* already referred to. Recollections of the event vary, but Zorch, early pioneers of synthesizer-driven progressive rock, seemingly played from a small stage facing the stones. After the solstice, around 30 people decided to stick around and set up a camp to continue the festival beside the by-way to the west of Stonehenge. Calling themselves the Wallies of Wessex, their open camp, known as Fort Wally, remained in the area until the winter solstice (Worthington 2004: 38–40). The following year a much bigger festival was planned, posters designed by Roger Hutchinson promoting the event being widely circulated across southern England (Fig. 8.6). The gathering was focused on King Barrow Ridge east of Stonehenge, and the hot sunny weather attracted a crowd of about 3000 to hear two bands, Hawkwind and Here and Now, open the show on Midsummer's Eve. The festival lasted about ten days and was generally judged to be a success. It was repeated in 1976, back to the west of Stonehenge, and continued there annually until 1984.

Of all the bands that performed at the Stonehenge festival, space-rockers Hawkwind will for ever be the best remembered, because of their regular performances there through the late 1970s and early 80s. Their last performance was recorded live and later released as the album *Stonehenge: This Is Hawkwind Do not Panic*, while material recorded at earlier performances was extensively used on the compilation album *Welcome to the Future* released in 1985. Some film footage of the festival also survives and has been released on DVD as *Solstice at Stonehenge*.

Over the course of a decade or so, the Stonehenge Free Festival became increasingly associated with a counter-culture that, while maintaining an interest in the stones, gradually drifted away from its hippie-dominated roots. Trouble was brewing, and in 1983 *The Stonehenge Regulations* were approved by parliament, the only such legislation relating to an ancient monument in Britain and still on the statute book, updated as Statutory Instrument 1997/2038. This would give legal authority to the desire to stop people gathering at or around the stones. The regulations were not immediately applied, but the anarchy of punk lifestyles, the rise of the peace movement, anti-Thatcherism protests, riots against the

Poll Tax, militant animal rights campaigns, attacks on the establishment, and the nihilistic culture of rave all came together as ingredients for unrest at the 1984 festival, an event claimed as the largest free music festival in the British Isles (Fig. 8.7). To many observers it was no surprise that English Heritage and the National Trust decided that enough was enough and banned the festival from taking place on their property in 1985. Equally, many others saw the move as provocative and a step too far in terms of restricting personal freedoms and controlling the will of counter-cultures.

Stonehenge and the Stonehenge festival became the focus of confrontation which erupted into violence on 2 June 1985 when Wiltshire Police stopped a convoy of would-be festival-goers heading for Stonehenge on the A338. Their convoy was chased into a field near Cholderton, where ensued what the press dubbed the Battle of the Beanfield between police and protestors (Worthington 2005). In true rock and roll style, the incident was taken as the theme for a rebellious song of the same name by indie rockers the Levellers (1991, track 10), and was also picked up by Roy Harper in the song 'Back to the Stones' eventually included amongst the live tracks on his album *Born in Captivity* (1994). The events surrounding the incident have been reported and analysed from a number of perspectives (for example, Worthington 2002), Chippindale concluding that most people find the beliefs of neo-Druids and New Age travelling communities more amusing than appalling, and that the festival's dreams were more eccentric than world-threatening (1986: 55).

In the aftermath, people shrewdly asked, 'Who owns Stonehenge?' (Chippindale *et al*), but little changed over the following years as English Heritage and the National Trust resolutely refused to allow not just the festival but any access to Stonehenge at all over the solstice period. Exclusion orders were imposed and a heavy police presence maintained year after year, things that riled those opposing the ban and provoked confrontation both peaceful and otherwise (Bender; Dobinson). Various proposals were made to provide a more worthwhile celebration (for example, Chippindale 1985), but it was not until 2000 that general access to the stones at the summer solstice was restored, a move that initially prompted a mixed reaction (Dennison). It took more than a decade for the anger and uptight attitudes of the 1980s to seep away, but in 2001 about 14,000 people turned up for what was a very peaceful and enjoyable night of celebrations; in subsequent years attendance has risen to around 20,000, with music of all kinds echoing around the stones through the night and well past dawn. Similar albeit smaller-scale celebrations at the site for the winter solstice and equinoxes have also sprung up.

Conclusion

Over the past 40 years or so Stonehenge has clearly had an impact on the popular music industry, especially the world of rock, although it is fair to say that when measured against the colossal scale of that industry, what is highlighted here is but a tiny slice. Even so, more people have probably come to know Stonehenge through promotional material for bands, albums and concerts than will ever encounter it in the pages of an academic study.

Looking at the origins of much of the material reviewed here, it is notable that there are two peaks in output. The first is from the mid 1960s through to the mid 1990s. The second started about 2005 and is still going. In large measure this second wave owes much to the power of the internet in providing limitless opportunities for musicians and artists to market and promote their own work. Much of what has appeared appeals to discrete subcultures (psychedelic rock, folk rock, indie, world music etc), perpetuating a tradition that can also be seen during the first wave of Stonehenge-inspired music in popular culture. Ironically, in a world dominated by rapidly changing technology for the creation, delivery, and enjoyment of music, Stonehenge remains an icon of choice for the perpetrators of rock music and their stooges, a source of inspiration and a rallying point for counter-cultures just as it has been for centuries.

ACKNOWLEDGEMENTS

I offer grateful thanks to Roger Bland, Vanessa Constant, Roger Doonan, David Gaimster, Chris Gerrard, Richard Morris, Neil Mortimer, Robert Orledge, Miles Russell, Yvette Staelens, Paul Stamper and Alan Saville for information about particular images, artists, composers, and designs, and for searching through their record collections to trace missing details.

FIGURE 8.6: POSTER DESIGNED BY ROGER HUTCHINSON ADVERTISING THE 1975 FREE FESTIVAL AT STONEHENGE

FIGURE 8.7: AERIAL VIEW OF STONEHENGE WITH THE 1984 FESTIVAL IN FULL SWING

LIST OF SOURCES

Books and articles

Anon: *The Illustrated Guide to Old Sarum and Stonehenge* (Salisbury, 1884)

——: *Lancaster's Stonehenge Hand-Book* (Salisbury, 1894)

Bailey, E: 'The notion of implicit religion: what it means, and does not mean', in *The Secular Quest for Meaning in Life: Denton papers in implicit religion* ed E Bailey (New York, 2002), 1–12

Barrett, J: *Fragments From Antiquity: an archaeology of social life in Britian, 2900–1200 BC* (Oxford, 1994)

Becker, J: *Deep Listeners: music, emotion, and trancing* (Bloomington IN, 2004)

Behrens, H: *Die Jungsteinzeit im Mittelelbe-Saale-Gebiet* (Berlin, 1973)

——: 'Neues und Altes zu den neolithischen Tontrommeln', *Fundberichte aus Hessen* 19–20: Festschrift U Fischer (1979), 145–61

—— and E Schröter: *Siedlungen und Gräber der TRB und Schnurkeramik bei Halle (Saale)* (Berlin, 1980). Veröffentlichungen des Landesmuseums für Vorgeschichte in Halle 34

Bender, B: *Stonehenge: making space* (Oxford, 1998)

Blacking, J: *How Musical is Man?* (Seattle, 1973)

——: 'Dance, conceptual thought and production in the archaeological record', in *Problems in Economic and Social Archaeology* ed G G Sieveking, I H Longworth and K E Wilson (London, 1976), 3–13

——: 'Reflections on the effectiveness of symbols', in *Music, Culture, Experience: selected papers of John Blacking* ed R Byron (Chicago, 1995), 174–97

Bloch, M: 'Symbols, song, dance and features of articulation, *or* Is religion an extreme form of traditional authority?' *Archives Européennes de Sociologie (European Journal of Sociology)* 15 (1974), 55–81

Blumenfeld, L: *Voices of Forgotten Worlds: traditional music of indigenous people* (Roslyn NY, 1993)

Bognár-Kutzián, I: *The Copper Age Cemetery of Tiszapolgár-Basatanya* (Budapest, 1963). Archaeologia Hungarica 42

Borsay, P: 'Sounding the town', *Urban History* 29 (2002), 92–102

Botha, R and C Knight eds: *The Prehistory of Language* (Oxford, 2009)

Brade, C: 'The prehistoric flute—did it exist?', *Galpin Society Journal* 35 (1982), 138–50

Bradley, R: *The Prehistory of Britain and Ireland* (Cambridge, 2007)

Brundel, A (Curator, Orkney Museum, Kirkwall): pers comm

Buckley, J *et al* eds: *Rock: The Rough Guide* (London, 2/1999)

Buisson, D: 'Les flûtes paléolithiques d'Isturitz (Pyrénées-Atlantiques)', *Bulletin de la Société préhistorique française* 87 (1990), 420–33

Burl, A: *The Stone Circles of Britain, Ireland and Brittany* (New Haven, 2000)

Camden, W, transl Philemon Holland: *Britannia* (London, 1610)

Catling, C: 'Message in the stones', *Current Archaeology* 212 (2007), 12–19

Chippindale, C: 'Time for a Stonehenge celebration', *Current Archaeology* 98 (1985), 84–5

——: 'Stoned Henge: events and issues at the summer solstice, 1985', *World Archaeology* 18 (1986), 38–58

——: *Stonehenge Observed: images from 1350 to 1987* (Southampton, 1987)

—— *et al*: *Who Owns Stonehenge?* (London, 1990)

——: *Stonehenge Complete* (London, 3/2004)

Christensen, C L: *Odeon Room Acoustic Program User Manual* (Lyngby, Denmark, 2008)

Clark, R: *The Multiple Natural Origins of Religion* (Bern, 2006)

Clarke, D V, T G Cowie and A Foxon: *Symbols of Power at the Time of Stonehenge* (Edinburgh, 1985)

Cleal, R M J, K E Walker and R Montague: *Stonehenge in its Landscape: twentieth-century excavations* (London, 1995). English Heritage Archaeological Report 10

Clodoré, T: 'De la préhistoire à l'âge du Bronze (9000–2300 avant J-C)' (*Préhistoire de la musique*: 47–57) [2002a]

——: 'L'introduction du métal dans le paysage sonore: de l'âge du Bronze à la fin premier âge du Fer (2300–450 avant J-C)' (*Préhistoire de la musique*: 59–99) [2002b]

Coles, J M and D D A Simpson eds: *Studies in Ancient Europe* (Leicester, 1968)

Conard, N J *et al*: 'Eine Mammutelfenbeinflöte aus dem Aurignacien des Geissenklösterle' *Archäologisches Korrespondenzblatt* 34 (2004), 447–62

Cornwell, B: *Stonehenge: a novel of 2000 BC* (London, 1999)

Crewdson, J: *Sounds of a Neolithic Landscape* (BSc honours thesis, London Guildhall University, 2002)

Cross, I and I Morley: 'The evolution of music: theories, definitions and the nature of the evidence', in *Communicative Musicality* ed S Malloch and C Trevarthen (Oxford, 2009), 61–82

—— and A Watson: 'Acoustics and the human experience of socially-organized sound' (Scarre and Lawson 2006: 107–116)

Cross, J: 'Birtwistle, Sir Harrison', *The New Grove Dictionary of Music and Musicians* ed S Sadie and J Tyrrell (London, R/2001), vol 3, 619–26

D'Errico, F and G Lawson: 'The sound paradox: how to assess the acoustic significance of archaeological evidence?' (Scarre and Lawson 2006: 41–57)

—— *et al*: 'Archaeological evidence for the emergence of language, symbolism, and music—an alternative multidisciplinary perspective', *Journal of World Prehistory* 17 (2003), 1–70

Dams, L: 'Preliminary findings at the "organ" sanctuary in the cave of Nerja, Malaga, Spain', *Oxford Journal of Archaeology* 3/1 (1984), 1–14

Darvill, T: 'Archaeology in rock', in *Material Engagements: studies in honour of Colin Renfrew* ed N Brodie and C Hills (Cambridge, 2004), 55–77 [2004a]

——: *Long Barrows of the Cotswolds and Surrounding Areas* (Stoud, 2004) [2004b]

——: *Stonehenge: the biography of a landscape* (Stroud, 2006)

Dauvois, M: 'Son et musique paléolithiques', *Les dossiers d'archéologie* 142 (1989), 2–11

Dennison, S: 'Archaeologists divide on Stonehenge solstice', *British Archaeology* 54 (2000), 4

Devereux, P: *Stone Age Soundtracks: the acoustic archaeology of ancient sites* (London, 2001)

—— and R Jahn: 'Preliminary investigations and cognitive considerations of the acoustical resonances of selected archaeological sites', *Antiquity* 70 (1996), 665–6

Dibble, J: *John Stainer: a life in music* (Woodbridge, 2007)

Dobinson, C: 'Saturday night and Sunday morning', *British Archaeological News* 7/4 (1992), Stonehenge supplement, 61–2

Donald, M: *Origins of the Modern Mind: three stages in the evolution of culture and cognition* (Cambridge MA, 1991)

Dronfield, J: 'Migraine, light and hallucinogens: the neurocognitive basis of Irish megalithic art', *Oxford Journal of Archaeology* 14 (1995), 261–75

——: 'The vision thing: diagnosis of endogenous derivation in abstract arts', *Current Anthropology* 37 (1996), 373–91

Dunbar, R: *The Human Story* (London, 2004)

Eason, J: *Conjectures on that Mysterious Monument of Ancient Art, Stonehenge* (London, 1815)

Ehrenreich, B: *Dancing in the Streets: a history of collective joy* (New York, 2007)

Fages, G and C Mourer-Chauviré: 'La flûte en os d'oiseau de la grotte sépulcrale de Veyreau (Aveyron) et inventaire des flûtes préhistoriques d'Europe', *Mémoires de la Société préhistorique française* 16 (1983), 95–103

Falk, D: 'Hominid brain evolution and the origins of music' (Wallin, Merker and Brown 1999: 197–216)

Feld, S: *Sound and Sentiment: birds, weeping, poetics, and song in Kaluli expression* (Philadelphia, 1982)

——: 'A poetics of place: ecological and aesthetic co-evolution in a Papua New Guinea rainforest community', in *Redefining Nature: ecology, culture and domestication* ed R F Ellen and K Fukui (Oxford, 1996), 61–87

Fischer, U: 'Zu den mitteldeutschen Trommeln', *Archaeologia Geographica* 2 (1951), 98–105

Forster, E M: *Collected Short Stories* (London, 1947)

Freeman, W: 'A neurobiological role of music in social bonding' (Wallin, Merker and Brown 1999: 411–22)

Gailli, R: *L'Aventure de l'os dans la préhistoire* (Nîmes, 1978)

Gell, A: 'The language of the forest: landscape and phonological iconism in Umeda', in *The Anthropology of Landscape: perspectives on space and place* ed E Hirsch and M O'Hanlon (Oxford, 1995), 232–54

——: *Art and Agency* (Oxford, 1998)

Gibson, E ed: *Camden's Britannia, Newly Translated into English* (London, 1695)

Gibson, J J: *The Senses Considered as Perceptual Systems* (London, 1966)

Giddens, A: *Modernity and Self-Identity* (Cambridge, 1991)

Gimbutas, M: *The Goddesses and Gods of Old Europe* (London, 1974, R/1982)

Goddard, E H: *Stonehenge Handbook and Guide* (Devizes, 1894)

Gorbman, C: *Unheard Melodies: narrative film music* (London, 1987)

Gordon, R: 'Cash for questions: Shaun Ryder', *Q* 176 (2001), 13–16

Grant, J, S Gorin and N Fleming: *The Archaeology Coursebook* (London, 2002)

Gray, J: *African Music* (London, 1991)

Hahn, J: 'Le Paléolithique supérieur de l'Allemagne méridionale (1991–1995)' in *Le Paléolithique supérieur européen: Bilan quinquennal (1991–1996)* ed M Otte (Forli, 1996), 181–6. ERAUL 76

Hardy, F E: *The Later Years of Thomas Hardy, 1892–1928* (London, 1930)

Hardy, T: *The Return of the Native* (London, 1878)

——: *The Trumpet-Major* (London, 1880)

——: *The Woodlanders* (London, 1887)

——: *Tess of the D'Urbervilles* (London, 1891)

——, M Millgate and F E Hardy: *The Life and Work of Thomas Hardy* (London, 1984)

Hargreaves, D J and A C North eds: *The Social Psychology of Music* (Oxford, 1997)

Harvey, G: *Animism: respecting the living world* (London, 2005)

Hawkins, G S: *Stonehenge Decoded* (London, 1966)

Herbage, J: booklet notes to Ireland 1966 (sound recording—see below), abridged 2007

Hickmann, E, A A Both and R Eichmann eds: *Musikarchäologie im Kontext: archäologische Befunde, historische Zusammenhänge, soziokulturelle Beziehungen* (Rahden, 2006). Studien zur Musikarchäologie 5

Hill, R: *Stonehenge* (London, 2008)

Hoare, R Colt: *The Ancient History of Wiltshire* (2 vols, London, 1812–21)

Holmes, P: 'The Scandinavian bronze lurs: accident or intent?' (Scarre and Lawson 2006: 59–69)

—— and J M Coles: 'Prehistoric brass instruments', *World Archaeology* 12 (1981), 280–86

Horton, P: *Samuel Sebastian Wesley: a life* (Oxford, 2004)

Howkins, A: 'The discovery of rural England', in *Englishness: politics and culture 1880–1920* ed R Colls and P Dodd (London, 1986), 62–88

Hutton, R: *The Druids: a history* (London, 2007)

——: *Blood and Mistletoe: the history of the Druids in Britain* (London, 2009)

Ingold, T: 'Stop, look and listen! Vision, hearing and human movement', *The Perception of the*

Environment: essays on livelihood, dwelling and skill (London, 2000), 243-87

Irwin, M: *Reading Hardy's Landscapes* (Basingstoke, 2000)

Jackson, A: 'Sound and ritual', *Man* 3 (1968), 293–9

James, H: *Plans and Photographs of Stonehenge* (Southampton, 1867)

James, M R: 'Casting the runes', *More Ghost Stories of an Antiquary* (London, 1911), 235–67

Jaynes, J: *The Origin of Consciousness in the Breakdown of the Bicameral Mind* (London, 2/1993)

Jones, L E: 'Everybody must get stoned: megaliths and movies', *3rd Stone* 40 (2001), 6–14

Kaufmann, D: 'Eine frühneolithische Gefäßrassel von Rossleben/Thüringen' (Hickmann, Both and Eichmann 2006: 97–108)

Kawada, J: 'Human dimensions in the sound universe', in *Redefining Nature: ecology, culture and domestication* ed R F Ellen and K Fukui (Oxford, 1996), 39–60

King, E: *Munimenta Antiqua* (London, 1799)

Kruth, P and H Stobart eds: *Sound* (Cambridge, 2000)

Kuttruff, H: *Room Acoustics* (London, 4/2000)

Lawson, A J: *Chalkland: an archaeology of Stonehenge and its region* (East Knoyle, 2007)

Lawson, G et al: 'Mounds, megaliths, music and mind: some thoughts on the acoustical properties and purposes of archaeological spaces', *Archaeological Reviews from Cambridge* 15 (1998), 111–34

Le Gonidec, M-B: 'Qu'est-ce-qu'un instrument de musique?' (*Préhistoire de la musique*: 23–31)

Leventhall, G: 'What is infrasound?', *Progress in Biophysics and Molecular Biology* 93 (2007), 130–37

Longmire, J: *John Ireland: portrait of a friend* (London, 1969)

Lund, C: 'The archaeomusicology of Scandinavia', *World Archaeology* 12 (1981), 246–65

——: 'Bone flutes in Västergötland, Sweden—finds and traditions', *Acta Musicologica* 57 (1985), 9–25

——: booklet notes, *Fornnordiska Klanger* (sound recording—see below)

Lynskey, D: 'This is for Romeo and Juliet', *Q* 179 (2001), 106–9

Maxfield, M: 'The journey of the drum', *ReVision* 16/4 (1994), 157–63

McLuhan, M: *The Gutenberg Galaxy: the making of typograhic man* (Toronto, 1962)

Megaw, J: 'Penny whistles and prehistory', *Antiquity* 34 (1960), 6–13

——: 'Problems and non-problems in palaeo-organology: a musical miscellany' (Coles and Simpson: 333–58) [1968]

——: 'The bone ?flute', in *Gwernvale and Penywyrlod: two Neolithic long cairns in the Black Mountains of Brecknock* ed W J Britnell and H N Savory (Cardiff, 1984), 27–8. Cambrian Archaeological Monographs 2

Merewether, J: *Diary of a Dean* (London, 1851)

Merriam, A P: *The Anthropology of Music* (Evanston IL, 1964)

Michael, W: *Stonehenge* (London, 1864)

Michell, J: *The View over Atlantis* (London, 1969)

Midgley, M S: *The Monumental Cemeteries of Prehistoric Europe* (Stroud, 2005)

Mildenberger, G: 'Die neolithischen Tontrommeln', *Jahresschrift Halle* 36 (1952), 30–41

——: 'Ein steinzeitlicher Grabhügel in der Harth (Kreis Leipzig)', *Arbeits- und Forschungsberichte zur sächsischen Bodendenkmalpflege* 3 (1953), 7–24

Milner, D ed: *The Highways and Byways of Britain* (London, 2008)

Mithen, S: *The Singing Neanderthals: the origins of music, language, mind and body* (London, 2005)

Montagu, J: 'The conch in prehistory: pottery, stone and natural' *World Archaeology* 12 (1981), 273–9

Morley, I: *The Evolutionary Origins and Archaeology of Music* (PhD thesis, University of Cambridge, 2003)

Morley, I: 'Hunter-gatherer music and its implications for identifying intentionality in the use of acoustic space' (Scarre and Lawson 2006: 95–105)

Müller, D W: 'Die Bernburger Kultur Mitteldeutschlands im Spiegel ihrer nichtmegalithischen Kollektivgräber', *Jahresschrift für mitteldeutsche Vorgeschichte* 76 (1994), 75–200

Müller, J: *Soziochronologische Studien zum Jung- und Spätneolithikum im Mittelelbe-Saale-Gebiet (4100–2700 v Chr)* (Rahden/Westf, 2001). Vorgeschichtliche Forschungen 21

Nettl, B: 'An ethnomusicologist contemplates universals in musical sound and musical culture' (Wallin, Merker and Brown 1999: 463–72)

Nitzschke, W: 'Eine verzierte Trommel der Salzmünder Kultur von Gerstewitz', *Ausgrabungen und Funde* 31 (1986), 149–51

Nketia, J H K: *The Music of Africa* (New York, 1974)

O'Dwyer, S: *Prehistoric Music of Ireland* (Stroud, 2004)

Orel, H: *Thomas Hardy's Personal Writings: prefaces, literary opinions, reminiscences* (Lawrence KS, 1966)

Parker Pearson, M: *The Archaeology of Death and Burial* (Stroud, 1999)

——: interview, *Stonehenge Decoded* (screen media production—see below)

—— et al: 'Materializing Stonehenge: the Stonehenge Riverside Project and new discoveries', *Journal of Material Culture* 11 (2006): 227–61

—— et al: 'The age of Stonehenge', *Antiquity* 81 (2007), 617–39

—— et al: 'Who was buried at Stonehenge?', *Antiquity* 83 (2009), 23–39

Piggott, S: *The West Kennet Long Barrow Excavations, 1955–56* (London, 1962). Ministry of Works Archaeological Reports 4

——: *The Druids* (London, 2/1975)

Pitts, M: *Hengeworld* (London, R/2001)

——: 'The big dig: Stonehenge', *British Archaeology* 102 (2008), 12–17

Pocock, D: 'The music of geography', in *Humanistic Approaches in Geography* ed D Pocock (Durham, 1988), 62–71. Durham University Department of Geography Occasional Publications (New Series) 22

——: 'Sound and the geographer', *Geography* 74 (1989), 193–200

Poe, E A: 'Ligeia', *The Complete Stories* (New York,

1992), 350–64

Pollard, J: 'The materialization of religious structures in the time of Stonehenge' *Material Religion* 5 (2009, forthcoming)

—— and A Reynolds: *Avebury: the biography of a landscape* (Stroud, 2002)

Pollex, A: 'Comments in the interpretation of the so-called cattle burials of Neolithic Central Europe', *Antiquity* 73 (1999), 542–50

Préhistoire de la musique: sons et instruments de musique des âges du Bronze et du Fer en France (Nemours, 2002)

Price-Williams, D and D J Hughes: 'Shamanism and altered states of consciousness' *Anthropology of Consciousness* 5/2 (1994), 1–15

Pryor, F: *Britain BC* (London, 2004)

Purser, J: booklet notes, *The Kilmartin Sessions* (sound recording—see below)

Ramachandran, V S: *Phantoms in the Brain* (London, 1999)

Reznikoff, I: 'The evidence of the use of sound resonance from Palaeolithic to medieval times' (Scarre and Lawson 2006: 77–84)

—— and M Dauvois: 'La dimension sonore des grottes ornées', *Bulletin de la Société Préhistorique Française* 85 (1988), 238–46

Richards, F: *The Music of John Ireland* (Aldershot, 2000)

Richards, J: *Stonehenge: the story so far* (Swindon, 2007)

Rouget, G: *Music and Trance* (Chicago, 1985)

Rumsey, F: 'Subjective assessment of the spatial attributes of reproduced sound', *Proceedings of the AES 15th International Conference: Audio, Acoustics and Small Space* (Copenhagen, 1998), 122–35

Rutherfurd, E: *Sarum* (London, 1987)

Scarre, C: 'Sound, place and space: towards an archaeology of acoustics' (Scarre and Lawson 2006: 1–10)

—— and G Lawson eds: *Archaeoacoustics* (Cambridge, 2006)

Schafer, R Murray: *Voices of Tyranny, Temples of Silence* (Indian River ON, 1993)

——: *The Soundscape: our sonic environment and the tuning of the world* (Rochester VT, 1977, repr 1994)

Sothern, P M T: 'A comparison of the medieval White Castle flute with the Chalcolithic example of Veyreau' *Proceedings of the Prehistoric Society* 55 (1989), 257–60

Seeger, A: 'Music and dance', in *Companion Encyclopedia of Anthropology* ed T Ingold (London, 1994), 686–705

Selkirk, A: 'Stonehenge bluestones: who is right?', *Current Archaeology* 226 (2009), 10

Sen, S N: *Acoustics, Waves and Oscillations* (New Delhi, 1990)

Skeates, R: 'Triton's trumpet: a Neolithic symbol in Italy', *Oxford Journal of Archaeology* 10 (1991), 17–31

Sloboda, J A: *The Musical Mind* (Oxford, 1985)

Smith, W: *A Smaller Classical Dictionary of Biography, Mythology and Geography* (London, 17/1877)

Stokes, M ed: *Ethnicity, Identity and Music: the musical construction of place* (Oxford, 1994)

Storr, A: *Music and the Mind* (London, 1992)

Stout, A: 'The world turned upside down: Stonehenge summer solstice before the hippies', *3rd Stone* 46 (2003), 38–42

Thom, A: *Megalithic Sites in Britain* (Oxford, 1967)

Thomas, J: 'The politics of vision and the archaeologies of landscape', in *Landscape: politics and perspectives* ed B Bender (Oxford, 1993), 19–48

Thorgerson, S and A Powell: *One Hundred Best Album Covers: the stories behind the sleeves* (London, 1999)

Tilley, C *et al*: 'Stonehenge: its landscape and its architecture', in *From Stonehenge to the Baltic: living with cultural diversity in the third millennium BC* ed M Larsson and M Parker Pearson (Oxford, 2007), 183–204. BAR International Series 1692

Treece, H: *The Golden Strangers* (London, 1956)

Tuan, Y-F: *Topophilia* (Englewood Cliffs NJ, 1974)

Turner, V: *The Ritual Process* (London, 1969)

Turow, G.: *Auditory Driving as a Ritual Technology* (Religious Studies honours thesis, Stanford University, 2005)

—— and J Berger eds: *Musical Time and Human Behavior: perspectives on rhythm in ritual and healing* (forthcoming)

Tuzin, D: 'Miraculous voices: the auditory experience of numinous objects', *Current Anthropology* 25 (1984), 579–96

Veenstra, A: 'The classification of the flute', *Galpin Society Journal* 17 (1964), 54–63

Vitebsky, P: *The Shaman* (London, 1995)

Wainwright, G J, with I H Longworth: *Durrington Walls: excavations 1966-68* (London, 1971)

Wallin, N L: *Biomusicology* (Stuyvesant NY, 1991)

——, B Merker and S Brown eds: *The Origins of Music* (Cambridge MA, 1999)

Watson, A: 'The sounds of transformation: acoustics, monuments and ritual in the British Neolithic', in *The Archaeology of Shamanism* ed N Price (London, 2001), 178–92 [2001a]

——: 'Composing Avebury', *World Archaeology* 33 (2001), 296–314 [2001b]

——: 'Making space for monuments: notes on the representation of experience', in *Substance, Memory, Display: archaeology and art* ed C Renfrew, C Gosden and E DeMarrais (Cambridge, 2004), 79–96

——: '(Un)intentional sound? Acoustics and Neolithic monuments' (Scarre and Lawson 2006: 11–22)

—— and D Keating: 'Architecture and sound: an acoustic analysis of megalithic monuments in prehistoric Britain', *Antiquity* 73 (1999), 325–36

—— ——: 'The architecture of sound in Neolithic Orkney', in *Neolithic Orkney in its European Context* ed A Ritchie (Cambridge, 2000), 259–63

—— and J Was: 'Monuments in concert: a journey through the landscapes of sound' (conference performance, Theoretical Archaeology Group, Cardiff, 1999)

Watt, R J and R L Ash: 'A psychological investigation of meaning in music', *Musicæ Scientiæ* 2 (1998), 33–54

Worthington, A: 'A brief history of the summer solstice at Stonehenge', *3rd Stone* 42 (2002), 41–7
——: *Stonehenge: celebration and subversion* (Loughborough, 2004)
—— ed: *The Battle of the Beanfield* (Teignmouth, 2005)
Wyatt, S: *The Drums of the Southern TRB* (PhD thesis, University of Edinburgh, 2007)
——: 'Whistling down the wind' (unpublished lecture, 2008)
——: 'Psychopomp and circumstance, or Shamanism in context', in *Studien zur Musikarchäologie* 7 ed E Hickmann, R Eichmann and L-C Koch (Rahden, forthcoming 2010) [2010a]
——: 'Die Zaubertrommel: symbol of transformation' (forthcoming 2010) [2010b]
Zillwood, F W: *Stonehenge* (London, 1855)
Zubrow, E B W and E C Blake: 'The origin of music and rhythm' (Scarre and Lawson 2006: 117–26)

Internet sites (checked for availability on 14 July 2009)

Absolute Astonomy: 'Maiden Castle, Dorset'
www.absoluteastronomy.com/topics/Maiden_Castle,_Dorset
Acoustics and Music of British Prehistory Research Network
ambpnetwork.wordpress.com
AkuTEK: 'Concert hall acoustics: parameters'
www.akutek.info/concert_hall_acoustics_files/parameters.htm
BBC Guernsey: 'Local history: Le Trépied'
www.bbc.co.uk/guernsey/content/articles/2008/10/03/le_trepied_dolmen_feature.shtml
CD Baby
www.cdbaby.com
Cope, J: 'Thomas Hardy: the shadow on the stone', *The Modern Antiquarian* (March 2005)
www.themodernantiquarian.com/forum
Devereux, P and J Wozencroft: 'A Stone Age Holy Land?', *The Landscape and Perception Project*
www.landscape-perception.com
Fischer, S von: 'Scenic and sonic structure: acoustics workshop' (2002)
e-collection.ethbib.ethz.ch/eserv/eth:26421/eth-26421-01.pdf
Gowen, M: 'Unique prehistoric musical instrument discovered in Co Wicklow' (May 2004, plus later updates)
www.mglarc.com/index.php
Harrington, R: 'The shadow of Stonehenge: paganism, fate and redemption in Thomas Hardy's *Tess of the D'Urbervilles*' (2006)
www.greycat.org/papers/tess.html
Jovcevska, J N D, transl A Vasilkova-Midoska: 'Globular flute: archaeological site Mramor near Cashka', *Republic of Macedonia Cultural Heritage Protection Office*
www.uzkn.gov.mk/dokumenti/flejta/GLOBULAR_FLUTE.pdf
Larsen, N W, E Olmos and A C Gade: 'Acoustics in halls for rock music' (conference paper, Joint Baltic-Nordic Acoustics Meeting, Mariehamn, Åland, 2004)
www.acoustics.hut.fi/asf/bnam04/webprosari/papers/o18.pdf
Maryhill model online (2005)
www.3dancientwonders.com
Maryhill Museum of Art
www.maryhillmuseum.org/about.html
Monumental
www.monumental.uk.com
O'Dwyer, S: 'Four voices of the Bronze Age horns of Ireland' (conference paper, 1st Symposium of the International Study Group on Music-Archaeology, Michaelstein, Germany, 1998)
homepage.eircom.net/~bronzeagehorns/documents/FourVoices.pdf
Skålevik, M: 'Room acoustic parameters and their distribution over concert hall seats' (AkuTEK paper, 2008)
www.akutek.info/Papers/MS_Parameters_Distribution.pdf
Soulodre, G A, M C Lavoie and S G Norcross: 'Temporal aspects of listener envelopment in multichannel surround systems' (AES Convention Paper 5803: 114th Convention, Amsterdam, 2003)
www.aes.org/e-lib/
Stanford Institute for Creativity and the Arts, Center for Arts, Science and Technology: 'Research' (2nd Symposium on Music, Rhythm and the Brain, 2007)
www.stanford.edu/group/brainwaves/2007/research.html
Tarasov, N: 'Die ältesten Flöten der Welt', *Windkanal* 2005/1, 6–11
www.windkanal.de/PDF/2005-1/Wika_2005_1_Aelteste-Floeten-der-Welt.pdf
Mama Lisa's World: England
www.mamalisa.com
Wiltshire Heritage
www.wiltshireheritage.org.uk
Wyatt, S: 'The classification of the clay drums of the southern TRB' (2008)
www.jungsteinzeit.de [2008c]

Screen media productions

Evil Aliens, director Jake West, music Richard Wells (UK, 2005)
Fiddlers Three, director Harry Watt, music Spike Hughes (UK, 1944)
Help!, director Richard Lester, music the Beatles and Ken Thorne (UK, 1965)
Muppets from Space, director Tim Hill, music Jamshied Sharifi (USA, 1999)
Night of the Demon, director Jacques Tourneur, music Clifton Parker (UK, 1957)
Quatermass, director Piers Haggard, music Nic Rowley and Marc Wilkinson (UK, TV mini-series, 1979)
Shanghai Knights, director David Dobkin, music Randy Edelman (USA, 2003)
Solstice at Stonehenge, Hawkwind (Cherry Red Films, 2004)
Stonehenge Decoded, director Christopher Spencer (National Geographic Channel, 2008)
Tess, director Roman Polanski, music Philippe Sarde, sound editing Hervé de Luze and Peter Horrocks,

sound effects Jean-Pierre Lelong (France/UK, 1979)
Tess of the d'Urbervilles, director Ian Sharp, music Alan Lisk (London Weekend Television, 1998)
——, director David Blair, music Robert Lane (BBC mini-series, 2008)
This is Spinal Tap (Embassy Pictures, 1984)
The Tomb of Ligeia, director Roger Corman, music Kenneth V Jones (UK, 1964)
Yan Tan Tethera, libretto Tony Harrison, music Harrison Birtwistle (Opera Factory, 1986)

Sound recordings

Aerosmith: *Rock in a Hard Place* (Columbia, 1982)
Andy Scott's Sweet: *A* (Jimco Records, 1995)
Black Sabbath: *Born Again* (Vertigo, 1983)
David Bacha Band: *No Sleep after Stonehenge* (Ozit, 2008)
The Disrupters: *Gas the Punx* (Overground Records, 2005)
The Dolmen: *Ah Ry Ah* (The Dolmen, 2008)
The Druids of Stonehenge: *The Druids of Stonehenge* (Sundazed [Vinyl EP], 2008)
Fornnordiska Klanger/The Sounds of Prehistoric Scandinavia (MSCD 101, Stockholm, 1991)
Gregory: *Circle of Time* (Soul Catcher Music, 2008)
Hail to the Stonehenge Gods: tribute to Black Sabbath (various artists, World War III, 2002)
Harper, Roy: *Born in Captivity II,* aka *Unhinged* (Griffin Music, 1994)
Havens, Richie: *Stonehenge* (Stormy Forest Productions, 1970)
Hawkwind: *Stonehenge: This Is Hawkwind Do not Panic* (SHARP, 1984, re-released in *Official Picture Logbook* [Flicknife, 1987] and on CD in *Zones/Stonehenge* [Flicknife, 1988])
——: *Welcome to the Future* (Mausoleum, 1985)
——: *Year 2000: Codename Hawkwind. Volume One* (New Millennium Communications, 1999)
Holland, Jools (and his Rhythm and Blues Orchestra): *Moving out to the Country* (Warner, 2006)
Ireland, John: 'The forgotten rite' and 'Mai-Dun' (Lyrita, 1966, re-released on CD as *Boult Conducts Ireland* [2007])
——: 'Legend' for piano and orchestra (*The Romantic Piano Concerto* 39, Hyperion CD, 2006)
Kellianna: *Lady Moon* (Kellianna, 2004)
——: *I Walk with the Goddess* (Kellianna, 2007)
The Kilmartin Sessions: the sounds of ancient Scotland (Kilmartin House Trust KHTCD1, 1997)
The Levellers: *Levelling the Land* (China Records, 1991)
Lloyd, George: *Iernin: an opera in three acts*, plus interview with Chris de Souza (Albany CD, 1994)
McCarty, James: *Out of the Dark* (High Octane, 1994)
Misty Celtic Morning (various artists, Uni/North Sounds, 2001)
Nijssen, Arend and Jere Vanloan: *Stonehenge* (Nijssen and Vanloan, 2007)
Pickard, John: 'Men of stone', *Gaia Symphony* (Doyen CD, 2005)
Ridley, Amie: *Crystology* (Amie Ridley, 2008)
Rock of Ages (various artists, Sanctuary, 2002)
Rolling Stones: *Their Satanic Majesties Request* (Decca, 1967)
Runestone: *Stonehenge* (New World Music, 1992)
Sill, Gary and Bob Day: *Stonehenge Is Burning* (Gary Sill, 2007)
Spinal Tap: *This is Spinal Tap* (Polydor, 1984)
Steely Dan: *Remastered: The Best of Steely Dan* (MCA, 1993)
Stonehenge: Mystic Circle (various artists, Delta, 2000)
Stonehenge: *Remember Me?* (Montefiore Enterprises, 2008)
Stray: *New Dawn* (Mystic Records, 1998)
Ten Years After: *Stonedhenge* (Decca, 1968, remastered CD release BGO Records, 1997)
Wheater, Tim: *Sound Medicine Man* (New World, 2004)
White, Basil: *Peeing on Stonehenge* (Basilwhite, 2005)
Yes: *Tales from Topographic Oceans* (Atlantic, 1973)
The Zombies: *Odessey* [sic] *and Oracle* (CBS, 1968)

www.ingramcontent.com/pod-product-compliance
Lightning Source LLC
Chambersburg PA
CBHW061548010526
44115CB00023B/2984